工程建筑学概论

Outline

of

Engineering-Integrated Architecture

李兴钢 著
Li Xinggang

中国建筑工业出版社
CHINA ARCHITECTURE & BUILDING PRESS

自序 [1]

[1] 参考郭屹民的发言。李兴钢，郭屹民，等. 建筑学中的工程、
技术与意匠 [J]. 当代建筑，2021（10）：7-8.

Preface

基于长期多样实践的反馈和自身文化传统研究的启示，近十年以来我提出并着力论述的"胜景几何"，是对于建筑中"理想生活空间营造"的一种理论性描述。《胜景几何论稿》中对于"几何"的描述是：意指建筑的本体性操作，空间、结构、形式、建造等以几何为基础的互动衍化，是人工性与物质性的；对于"胜景"的描述是：意指一种与自然交互并紧密相关、不可或缺的空间诗性感知与体验，是自然性与精神性的；对于"胜景几何"的描述是：以结构、空间、形式、建造、基本单体等以几何为基础互动衍化的建筑本体元素，营造与自然交互并紧密相关的空间诗意，亦即营造人

Based on the feedback of long-term diverse practices and the insights of research on our own cultural traditions, the "Integrated Geometry and Poetic Scenery" that I have proposed and focused on in the past decade is a theoretical description of the "ideal living space creation" in architecture. The description of "Geometry" in the *Eassys about Integrated Geometry and Poetic Scenery* is: the ontological operation of architecture, the interactive derivation of space, structure, form, construction etc. based on the geometry, which is related to artificiality and materiality; The description of "Poetic" is: a kind of spatial poetic perception and experience that interacts with and is closely related to nature and indispensable, which is natural and spiritual.The description for "Integrated Geometry and Poetic Scenery" is: the architectural ontology elements interactively evoluted based on geometry, such as the structure, space, form, construction, basic monomer etc., are used to create a spatial poetry that interacts with and is closely related to nature, that is

工与自然之间的互成情境，它们所构成的整体，成为人（使用者和体验者）的理想生活空间。并提出：人，是由"胜景几何"所营造的"现实理想空间"中最为关键的核心主体。

　　当我拿着《胜景几何论稿》向陈军院士请教时，他脱口而出："'几何'可计算，'胜景'可模拟。"并鼓励和建议我可沿此方向深化研究。陈院士是航天测绘与地理信息领域的科学家，他应是非常敏感于一种理论成果的科学化描述和探究。建筑设计中空间、结构、形式、建造等的物质性的互动衍化是如何进行的？建筑设计中与自然交互并紧密相关、不可或缺的精神性的空间诗性感知与体验是怎样

to create an interactive situation between the artificial and the natural, and the whole they constitute becomes an ideal living space for people (users and experiencers). Furthermore: the people, are the most critical core subject in the "realistic ideal space" created by "Integrated Geometry and Poetic Scenery".

When I took a copy of *Eassys about Integrated Geometry and Poetic Scenery* and asked Academician Chen Jun for advice, he blurted out: "the 'Integrated Geometry' can be calculated, and the 'Poetic Scenery' can be simulated", and encouraged and suggested that I could further my research in this direction. Academician Chen, a scientist in the field of aerospace surveying and geographic information, should be susceptible to the scientific description and exploration of a theoretical achievement. How should the materialistic interactive derivation of space, structure, form, construction, etc. be carried out in architectural design? How is the spiritual spatial

被营造的？"现实理想空间"中最为关键的核心主体——人，获得了什么样的使用和体验？这一切都值得被精确描述和深度探究。

"万物有道"，道即秩序，而"万物皆数"。在路易·康的"静谧与光明"理念中，尚未存在的、不可度量的事物为"静谧"，已经存在的、可度量的事物为"光明"。建筑师的工作应该始于对不可度量的领悟，经由可度量的手段、工具设计和建造，最后完成的建筑物能生发出不可度量的气质，将我们带回最初的领悟之中。

古往今来，建筑一直被视为技术与艺术融合的产物，

poetic perception and experience that interacts with and is closely related to nature and indispensable in architectural design created? What kind of use and experience does the most critical core subject in the "realistic ideal space" – people-get? All of these deserve to be precisely described and explored in depth.

"Everything has Tao", Tao is order, and "everything is number". In Louis Kahn's concept of "Between Silence and Light", things that do not yet exist and that cannot be measured are "silence", and things that already exist and that can be measured are "light". The architect's work should begin with an understanding of the immeasurable, being designed and constructed by measurable means and tools, and the eventual finished building can generate an immeasurable temperament that brings us back to the original understanding.

Throughout the ages, architecture has been regarded as a product of the

而具有"动态有序"特征的复杂性科学延伸至当代建筑学领域，体现为与空间环境关系密切、设计路径多维因素共存、技术工具高阶高效的"复杂性建筑"，亟待构建适应已发生巨变的城乡建成空间环境的新理论、新方法、新技术和新工具。"工程建筑学"在这样的背景下应运而生。如果说"胜景几何"是针对复杂性建筑的设计理论和范式，"工程建筑学"即是针对复杂性建筑的系统化设计方法和手段。在工程原理和技术场景的引导之下，工程建筑学设计方法支撑了胜景几何建筑设计理论。以胜景几何理论和工程建筑学方法为核心，构建起一种当代的复杂性建筑设计理论、

fusion of technology and art, and complexity science with its characteristics of "dynamic order" extends to the field of contemporary architecture, where it is embodied in a "complexity architecture" that is closely related to the spatial environment, with multi-dimensional factors coexisting in the design approach and supported by advanced and efficient technical tools. It is therefore urgent to build new theories, new methods, new technologies and new tools to adapt to the urban and rural built-up spatial environment that has undergone significant changes. "Engineering-Integrated Architecture" came into being in this context. If "Integrated Geometry and Poetic Scenery" is a design theory and paradigm for complexity architecture, " Engineering-Integrated Architecture" is then a systematic design method and means for complexity architecture. Under the guidance of engineering principles and technical scenarios, the Engineering-Integrated Architecture design method supports the architectural design theory of Integrated Geometry and Poetic Scenery. With the theory of Integrated Geometry and Poetic Scenery and the method of Engineering-Integrated Architecture as the core, a contemporary

方法和技术体系。

　　本书作为一种框架性的"概论"，述及了与工程建筑学相关的研究背景、理论、方法、技术概要和对应的工程应用实践案例，尚需深化、细化、成熟，也仍在不断思考和完善之中，期待大家的批评指正。

　　工程建筑学的提出，鼓励和强调技术、工程、科学与设计、艺术在建筑中的高度融合，但同时特别要重视的一点是："工程或技术需要以人类学的关怀来呈现，才能真正属于建筑学。"[1] 好建筑最终仍然应是一种能够给我们带来"不可度量"之领悟的事物，它不是一种精密获取的计算结

[1] 参考郭屹民的发言。李兴钢，郭屹民，等. 建筑学中的工程、技术与意匠 [J]. 当代建筑，2021（10）：7-8.

complexity architectural design theory, method, and technical system is constructed.

As a framework "introduction", this book describes the research background, the outline of theory, method and technology and the corresponding engineering application cases related to Engineering-Integrated Architecture. It still needs to be deepened, refined, and fully developed and is constantly being rethought and improved. The author looks forward to everyone's criticism and correction.

The proposal of Engineering-Integrated Architecture encourages and emphasizes a high level of integration of technology, engineering, science and design, art in architecture, but at the same time, special attention should be paid to: "Engineering or technology needs to be presented with anthropological care in order to truly belong to architecture"[1]. Eventually, good architecture should still be something that can give us an understanding

果，它不应失去"模糊"和"测不准"的特性，因为那也是有关建筑的人性体验中不可或缺的部分，"我们需要在充满着机智和幽默的、与人类智慧相关的'巧妙'建筑中获得人类所具有的共通情感——喜怒哀乐。"[1] 如何能、是否能真正"精确"地"模拟"这一"胜景"，对于工程建筑学而言，或许才是真正的终极挑战。

李兴钢
2022 年 5 月于北京

[1] 参考郭屹民的发言。李兴钢，郭屹民，等. 建筑学中的工程、技术与意匠 [J]. 当代建筑，2021（10）：7-8。

of the "immeasurable". It is not an acquired precise calculation result, and should not lose the "fuzzy" and "inaccurate" characteristics, because it is also an indispensable part of the human experience about architecture. "We need to acquire the common emotion that human beings have in 'ingenious' architecture that is full of wit and humor, and related to human intelligence – joy, anger, sorrow and happiness."[1] If and how we can truly "accurately" "simulate" this "poetic scenery" may be the real ultimate challenge for Engineering-Integrated Architecture.

Li Xinggang
May 2022 in Beijing

目 录

Contents

导言

由建筑学科困境
引出的思考

Introduction

Reflections

on

the Dilemma of Architecture

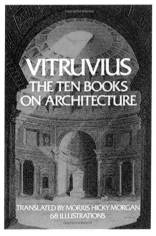

图 0-1

[1] 选自雷姆·库哈斯在 2000 年普利茨克奖颁奖礼上的获奖感言。
From Rem Koolhaas' acceptance speech at the 2000 Pritzker Prize ceremony.

图 0-1 维特鲁威《建筑十书》
Fig. 0-1 Vitruvius *The Ten Books on Architecture*

一、建筑学的危机和困境

雷姆·库哈斯（Rem Koohaas）曾说过这样一句话："我们仍沉浸在砂浆的死海中。如果我们不能将我们自身从永恒中解放出来，转而思考更急迫、更当下的新问题，建筑学不会持续到 2050 年。"[1] 建筑学将危机至此？当时我们的感觉是他在危言耸听，而 21 年后的今天，却越发感同身受，感触深刻。

延续了两千多年的维特鲁威的建筑三原则——坚固、适用、美观（图 0-1），所对应的人类生存条件或空间环境

I. Crisis and dilemma of architecture

"Only we architects don't benefit from this redefinition marooned in our own Dead Sea of mortar. Unless we break our dependency on the real and recognized architecture as a way of thinking about all issues, from the most political to the most practical, liberate ourselves from eternity to speculate about compelling and immediate new issues, such as poverty, the disappearance of nature, architecture will maybe not make the year two-thousand-fifty." [1]said Rem Koolhaas. Is architecture trapped in such a desperate situation? At that time, we might feel that he was intimidating us, but today, 21 years later, the sympathy is getting stronger.

Vitruvius's three principles of architecture—firmitas, utilitas, and venustas ("solidity," "utility," and "beauty") have lasted for more than two thousand years (Fig. 0-1). While for a long time, the corresponding human living

很长时间内是相对稳定的荒野自然或人文自然，对应的案例分别为马丘比丘城和古罗马城。但最近百余年来，人类的活动，特别是城镇化，使得人类所赖以生存的空间环境发生了很大变化（图0-2）。

从制图术（尺规）到数字技术（计算机），再到人工智能与虚拟现实技术，都是传统建筑学在自闭系统内进行的工具化推进或专业内的自说自话。当代建筑学和建筑设计有两个方向的发展趋势和现象值得关注：一是对于艺术创作属性的过分发挥，导致了主观随意的突出个人化表达的建筑；二是对于科学技术属性的过度强调，导致刻板冷

图 0-2 巨变的城镇化空间的图示表达
Fig. 0-2 Changing urbanization diagrammatic presentation

conditions or "spatial environment" were relatively stable wild nature or humanized nature, as shown in Machu Picchu and ancient Rome. Instead, in the last hundred years, human activities, especially urbanization, have significantly changed the environment of human living space (Fig. 0-2) .

From cartography (using a ruler) to digital technology (using a computer) to artificial intelligence and virtual reality technology, all these promotions are only advances in instruments, or rather the professional internal monologue in the self-enclosed systems of traditional architecture. However, the two trends and phenomena are remarkable in contemporary architecture and architectural design field: The excessive development of the attribute of "artistic creation" leads to a subjective or arbitrary architecture that highlights individual expressions; the overemphasis on the attribute of "science and technology" leads to a stereotyped and indifferent architecture

图 0-2

漠的与人的生活和情感疏离的建筑。这些现象的发生，与以下两种状况密切相关：经典建筑学的理论、方法和实践，无法全面适应人类生活多元环境（生态、技术、空间）的巨变并与之密切互动；同时，现代科技的发展和工程技术学科的独立（过度专业化）造就了建筑与工程严重分离的状况。这是导致建筑学茫然失措的学科困境乃至行业危机的根源。

二、失态、失序、失衡

城镇化快速发展，带来当代城乡建成环境的三大问

that is alienated from human living and emotions. The occurrence of these phenomena is closely related to the following two conditions: The classical theory, method, and practice of architecture cannot completely adapt to, or closely interact with, the dramatic changes in the diversified human living environment (ecological, technological and spatial). At the same time, both the development of modern science and technology and the independence (over-specialization) of engineering technology disciplines have created a severe separation of architecture and engineering. All of these are the ultimate reasons for the dilemma of architecture, the predicament of the discipline, and even the crisis of the whole industry.

II. Discord, Disorder, Imbalance

The rapid development of urbanization has brought about three major issues to the contemporary urban and rural environment: the "Discord" of the

题：生态环境的"失态"，技术环境的"失序"，空间环境的"失衡"。

"失态"是指超级或大型建筑对城市生态的排斥和对自然生态（包括气候环境）的破坏，例如平壤柳京大饭店和贵州某县城综合体。

"失序"是指随意的形式追求，技术与艺术脱离等造成技术环境（包括卫生环境）的失序，"奇怪"建筑频出，如河北"福禄寿"酒店。

"失衡"是指自然（原生自然、文明自然）环境（包括社会经济环境）的过度人工化和两极分化，如孟买城市中

ecological environment, the "Disorder" of the technical environment, and the "Imbalance" of the spatial environment. "Discord" refers to the exclusion of urban ecology and the destruction of natural ecology (including climate environment) by gigantic buildings, such as the Ryugyong Hotel in Pyongyang and a county complex in Guizhou.

"Disorder" refers to the phenomenon of disorder in the technical environment (including the sanitary environment) caused by the blind pursuit of forms and the separation between technology and art, resulting in the frequent appearance of "strange" buildings, such as the "Fu Lu Shou" Hotel in Hebei.

"Imbalance" refers to the over-artificialization and polarization of the natural (primitive and civilized) environment (including the socio-economic environment), such as the central area of Mumbai and the demolished Kowloon Walled City in Hong Kong (separated by a bay from the affluent area of Hong Kong Island).

心区和香港九龙城寨（与港岛的半山富人区一湾之隔，现已被拆除）。

三、生态、有序、平衡

"失态、失序、失衡"是城乡建成环境领域公认的世界级难题，对当下乃至未来的中国尤为紧迫和严峻。长期以来国内外各种建筑流派层出不穷，都无法全面回应和解决这一学科和行业的困境，回归和发展建筑与工程高度融合、触发、统合的一体化思想，建立和发展更具系统性、科学性、多元多变空间环境适应性及中国特色的建筑设计及其

III. Concord, Order, Balance

"Discord, Disorder, Imbalance" are recognized world-class problems in the urban and rural built environment. Furthermore, they are particularly urgent and severe for China today and in the future. Various domestic and foreign architectural schools have been emerging for a long time. However, none of them can comprehensively answer or solve this dilemma of the discipline and industry. Therefore, it is crucial to the architectural field and urgent to the country's high-quality urbanization development to backtrack and develop the idea of merging, triggering, and integrating in architecture and engineering while establishing and developing a methodology of architectural design and theoretical method that is more systematic, scientific, and adaptable to the diversified and variable spatial environment with Chinese characteristics.

理论方法体系，是建筑领域乃至国家高质量城镇化发展的迫切需要。

　　笔者经过长期的建筑设计及其理论研究与实践，谨慎思考并尝试提出的工程建筑学设计方法体系，在建立"胜景几何"建筑设计理论与范式的基础上，创新生态—技术—环境引导的建筑设计方法，研发基于可持续目标的系列关键技术，努力引领、回归和达到当代空间环境的生态、有序、平衡。

After long-term architectural practice and theoretical research, based on the establishment of the architectural design theory and paradigm of "Architecture Interacting with Nature – Integrated Geometry and Poetic Scenery", the author has carefully contemplated and tried to propose methodology of Engineering-Integrated Architecture, an innovative ecological-technical-environment-guided architectural design method, and has been researching and developing a series of key technologies based on sustainable goals, striving to lead, return to and achieve the "Concord, Order, Balance" of the contemporary spatial environment.

第一章

工程建筑学

Chapter I

Engineering-Integrated

Architecture

图 1-1

一、从结构建筑学到工程建筑学

1999 年建筑师赫尔穆特·杨（Helmut Jahn）与结构工程师维尔纳·索贝克（Werner Sobek）共同出版的作品集 *Archi-Neering*（图 1-1），首次以合成词的方式表述建筑与工程的结合。2014 年 10 月，同济大学关于同一主题的中日学术研讨会将其译为"结构建筑学"[1]。日本"Archi-Neering Design"模型展的发起人和策划者斋藤公男先生则译之为"建筑创新工学"[2]（图 1-2）。无论何译，都是希望改变现代科技发展和工程技术学科独立所造成的建筑与

[1] 钱锋，余中奇. 结构建筑学——触发本体创新的建筑思维 [J]. 建筑师，2015（2）：26.
[2] 斋藤公男. 建筑学的另一种视角——"何为建筑创新工学" [J]. 王西，译. 建筑师. 2015（2）：13.

图 1-1 *Archi-Neering*
Fig. 1-1 *Archi-Neering*
图 1-2 《建筑结构创新工学》
Fig. 1-2 *Archi-Neering Design*

I. From Archi-Neering to Engineering-Integrated Architecture

In 1999, architect Helmut Jahn and structural engineer Werner Sobek published a collaborative collection of works named Archi-Neering (Fig. 1-1), which was the first compound word to express the combination of architecture and engineering. In October 2014, Tongji University's Sino-Japanese Academic Symposium was on the same topic and translated it as "结构建筑学"[1](Structure-Integrated Architecture). Then, the initiator and curator of Japan's "Archi-Neering Design" model exhibition, Mr. Saito Koo, translated it as "Architectural Innovated Engineering"[2](Fig. 1-2). All these adoptions, no matter how translated, aim to change the condition caused by the development of modern science and the separation of engineering technologies, in which architecture and structure/engineering are separated into different disciplines. They express a strong desire to

图 1-2

结构 / 工程分离的状况，表达一种探讨结构 / 工程问题如何 "积极转化为建筑学思考的强烈意愿"，倡导一种由结构 / 工程构思产生建筑空间和形态特征的设计方式。

工程建筑学（Engineering-Integrated Architecture），意为工程原理和技术场景引导的建筑设计，即在关于建筑乃至空间环境的思考、设计和实施中，建造活动所包含的各类工程路径影响建筑空间和形式等本体特征产生的建筑创作思想，倡导 "以工程引导意匠" "将技术转化为诗意"。工程建筑学的提出，既是对建筑意匠和诗学传统与工程原理和技术场景高度交互、触发、统合的一体化思想的延续，

explore how structural/engineering issues can be "positively transformed into architectural thinking" and advocate a design approach: generating architectural space and morphological features from structural/engineering perspectives.

Engineering-Integrated Architecture means architectural design guided by engineering principles and technical conditions. Specifically, in the deliberation, design and creation of architecture and spatial environment, various engineering approaches in the construction can influence the architectural creation ideas generated by ontological characteristics, including architectural spaces and forms, while advocating "engineering-led ingenuity" and "technology transformed to poetics". The proposal of Engineering-Integrated Architecture is not only a continuous integration of highly interacting, triggering, and uniting architectural design and

也是在当代条件下，从 *Archi-Neering* 或结构建筑学强调结构对建筑设计作用，向更为全面的工程、技术对建筑设计作用的修正和发展。

二、工程建筑学的多元"三位一体"

工程建筑学体现科学思维、工程思维、技术思维、艺术思维在建筑设计中的高度统一，强调"技术—哲学—实践"三位一体、"工程—科学—艺术"三位一体、"空间—技术—形态"三位一体、"生态—技术—环境"三位一体（图1-3）。

poetic traditions with engineering principles and technical scenarios, but also a revision and evolution of Archi-Neering, or Structure-Integrated Architecture, which emphasizes the role of structure in architectural design, to a more comprehensive role of engineering and technology in architectural design under current conditions.

II. The diversified "Trinity" of Engineering-Integrated Architecture

Engineering-Integrated Architecture embodies the high unity of scientific thinking, engineering thinking, technical thinking, and artistic thinking in architectural design, emphasizing the trinity of "technology-philosophy-practice", the trinity of "engineering-science-art", the trinity of "space-technology-form", and the trinity of "ecology-technology-environment" (Fig. 1-3).

三、工程建筑学设计方法体系

　　工程建筑学设计方法体系包括："胜景几何"建筑设计理论与范式，"生态工程原理引导意匠创造""技术应用场景引导审美创作""遗产基因机制引导空间胜景"等工程建筑学设计方法和关键技术（图1-4）。

图 1-3 工程建筑学的多元"三位一体"
Fig. 1-3 The Diversified "Trinity" of Engineering-Integrated Architecture

III. Design methodology of Engineering-Integrated Architecture

Design methodology of Engineering-Integrated Architecture contains: the architectural design theory and paradigm of "Integrated geometry and poetic scenery", the architectural design methodology of "The principle of ecological engineering guides ingenious invention" "Technical application scenario guides aesthetic creation" "Inherited gene mechanism guides poetic scenery of space", etc (Fig. 1-4).

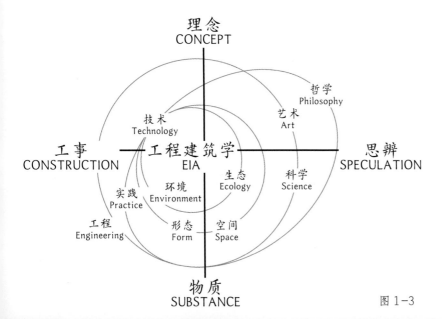

图 1-3

"胜景几何"
建筑设计理论与范式
"Integrated Geometry
and Poetic Scenery"

"胜景几何" 建筑设计理论
Design Theory of "Integrated Geometry and Poetic Scenery"

现实理想空间营造范式
Paradigm for Realistic Ideal Spatial Creation

面向现实的理论范式性实践
Paradigmatic Practice of Reality-Oriented Theory

工程建筑学
Engineering-Integrated
Architecture

生态工程原理引导意匠创造
Ecological Engineering Principle Guides Ingenious Invention

技术应用场景引导审美创作
Technical Application Scenario Guides Aesthetic Creation

遗产基因机制引导空间胜景
Inherited Genetic Mechanism Guides Poetic Scenery of Space

图1-4 工程建筑学设计
方法体系
Fig. 1-4 Design methodology
of Engineering-Integrated
Architecture

五点要素
Five Elements

五种策略
Five Strategies

十种模式 ─────────────────────────── 元上都遗址博物馆
Ten Methods   元上都遗址工作站
                                               绩溪博物馆
──────────────────────────────────── 安仁大匠之门文化中心

基于空间耦合的多维空间几何构型 ──────────→ 北京 2008 年奥运会主体育场暨
Multidimensional spatial geometry based on spatial coupling   国家体育场（"鸟巢"）

基于生态诱导的大型场馆设施与自然环境协调共生 ─────→ 北京 2022 年冬奥会与冬残奥会
Coexistence of the facilities of large venue and natural environment based on   延庆赛区
ecological induction

基于双碳目标的自能利用、节流开源 ───────── 天津大学新校区综合体育馆
Using self-generated energy, Reducing consumption and increasing gains to   延庆山地新闻中心
realize the "double carbon" target

数字智慧技术融合场地设计和赛道生成及人体运动 ─────→ 国家雪车雪橇中心
Digital intelligent technology integrated site design, track generation, and body movement   海南国际会展中心
                                               玉环博物馆和图书馆
结构及建造技术引导空间生成和形式意匠 ────────── 上海临港星空之境公园日月桥和
Structure and construction technology guide spatial generation and form design   无限桥

功能工艺技术融合城市空间和景观营造 ────────── 北京西直门交通枢纽
Integrating urban space and landscape creation with functional and process technology   北京地铁昌平线西二旗站
                                               北京朝阳区垃圾焚烧发电中心
环境调控技术引导建筑空间和景观营造 ────────── 威海名座大厦
Environmental control technology guides architectural space and landscape creation   中国驻西班牙大使馆办公楼改造

历史遗产"新旧相生、长效利用" ───────────── 首钢工舍
"Coexistence of the old and new with long-term utilization" for historic heritage   四川廖维公馆改造暨安仁古镇
                                               游客中心
旧城疏解改造"分形加密、重建规制" ─────────→ 北京大院胡同 28 号改造
"Type encryption and reconstruction" to relieve and transform old urban contexts

居住建筑"理想空间、当代重构" ───────────── 唐山第三空间综合体
"Ideal space and contemporary reconstruction" for residential architecture   成都安仁里居住小镇

乡村建造"空间记忆、在地建造" ───────────→ 楼纳露营基地服务中心
"Spatial memory and local identity" for rural development

图 1-4   27

第二章

"胜景几何"
建筑设计
理论与范式

Chapter II

Architectural Design
Theory and Paradigm
of
"Integrated Geometry and Poetic Scenery"

图 2-1

[1] 李兴钢.胜景几何论稿[M].
杭州：浙江摄影出版社，
2020:412.
[2] 殷瑞钰，汪应洛，李伯聪，
等.工程哲学[M].3版.北
京：高等教育出版社，2018:20.

图 2-1《胜景几何论稿》
Fig. 2-1 *Eassys about Integrated Geometry and Poetic Scenery*

一、"胜景几何"建筑设计理论

"胜景几何"建筑设计理论[1]，基于对传统和当代城市、聚落及建筑的田野考察、系统研究和实践反馈，将建筑学与工程哲学[2]、人文地理学、风景园林学、环境心理学等学科交叉融合，深度分析和揭示了复杂空间环境的类别及其与建筑之间的作用机制，将多元的空间环境分解为对应"原生自然"的自然空间环境和对应"人工自然"的城市空间环境，把"自然"要素纳入建筑本体，将建筑设计中的"几何"与"胜景"——亦即人工与自然、工程与意匠、本体与情境互动相成。"强调人工交互自然"的"胜景几何"建筑设计理论，是将中国的"天人合一"传统、复杂性科学

I. Architectural design theory of "Integrated Geometry and Poetic Scenery"

Architectural Design Theory of "Integrated Geometry and Poetic Scenery"[1], based on field investigation, systematic research and practical feedback on traditional and contemporary cities, settlements, and buildings, combines architecture with the philosophy of engineering[2], anthropogeography, landscape architecture, environmental psychology, and other disciplines, and deeply analyzes and reveals the types of complex spatial environments and their interaction mechanisms with buildings. It decomposes the diverse spatial environment into a natural space to "original nature" and urban space to "artificial nature", incorporating the element of "nature" into the building itself, resulting into an interaction and mutual promotion between design, be also known as the artificial and the natural, engineering and ingenuity, ontology and circumstances. The architectural design theory of "Integrated Geometry and Poetic Scenery" is comprehensive application of traditional

研究和当代技术的"交互"理念综合运用于建筑领域，并使之适用于复杂多元的当代人居环境和社会发展（图2-1）。

二、"现实理想空间营造范式"

建立"现实理想空间营造范式"[1]及其五点要素、五个策略、十种设计操作模式（图2-2），形成既延续传统，又有当代创新性的建筑设计范式体系，为再造建成空间环境"生态、有序、平衡"的人居胜景提供具系统性和科学性的建筑设计路径，与绿色"双碳"目标、增量有序发展、存量更新进化等当代建设需求高度契合。

五点要素：环境形势、人作天工、结构场域、叙事空

[1] 李兴钢. 胜景几何论稿[M]. 杭州：浙江摄影出版社，2020:105.

China's "harmony between universe and human", complexity of scientific research and the concept of "interaction" in contemporary technology to the field of architecture, and to make it applicable to the complex and diverse human inhabited environment and social development (Fig. 2-1).

II. "Paradigm for realistic ideal spatial creation"

The "Paradigm for realistic ideal space creation"[1] has been established with its five elements , five strategies, and ten design manipulating methods (Fig. 2-2), forming the methodology of architectural design that not only inherits the traditional procedure but also comprises contemporary innovation to provide a systematic and scientific architectural design method for the reconstruction of human settlements with a "Concordant, Orderly, Balanced" spatial environment, highly consisting with contemporary construction needs such as green "carbon peaking and carbon neutrality" goals, orderly

图 2-2 "现实理想空间营造范式" 及其五点要素、五个策略、十种设计操作模式
Fig. 2-2 "Paradigm for realistic ideal space creation" and its five elements, five strategies, and ten design manipulating methods

"理想空间"的选址、布局与所在自然环境（地理、气候、地质、人文、历史等条件乃至建成环境）的综合分析判断密切关联、相辅相成，并关联于场所感的营造，其中主体空间的坐落与方位、朝向至关重要；远势近形，"形"与"势"相辅相成、相互转化，并予人以动静不同、丰富生动的视觉感受。

The site selection and layout of the "ideal space" are closely related to the comprehensive analysis and judgment of the "natural" environment (geography, climate, geology, humanities, history and other conditions, even the built environment), and are related to the creation of a sense of place. Among them, the location, orientation and direction of the main space are crucial. Shape and potential complement each other, transform each other, and give people different dynamic and static, rich and vivid visual experiences.

人造的实物结构与多种自然要素的高度交互结合，互动互成。人工与自然模糊界限，相互衍生，转化为"自然化的人工"和"人工化的自然"，并升华为环境中新的生命"造化"。城市、乡村、建筑、园庭的一体化结构与全地景性聚落营造。

The artificial physical structure is highly combined with the interaction of a variety of natural elements, and their interaction is mutual and reciprocal. The boundaries between the artificial and the natural are blurred, and they derive from each other, transform into the "naturalized artificial" and the "artificial nature", which sublimate into new life "creation" in the environment, becoming an integrated structure of cities, villages, buildings and gardens and a creation of full-landscape settlements.

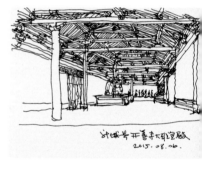

建筑的结构、空间、形式之间基于几何逻辑和人（使用者或体验者）的身体状态互动、衍化与匹配，营造建筑的独特气质、精神和动人的空间氛围；结构/空间单元水平或垂直的分解组合，明暗、虚实、大小、高下等建筑及空间意匠的推敲、营造与呈现。

The interaction, derivation and matching among the structures, spaces and forms of the building are based on geometric logic and the physical state of the person (user or experiencer), creating a unique temperament, spirit and impressive spatial atmosphere of the building; The horizontal or vertical decomposition and combination of structural/spatial units is deliberated, created and presented through architectural and spatial ingenuity such as contrasts of light and dark, void and solid, great and small, high and low, etc.

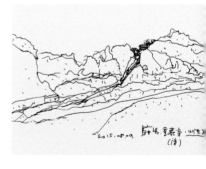

"疏密得宜、曲折尽致"——对于空间和景象的深远动人之体验，被不断串联起来并层层展现出来；以空间的方式，引导人经历"外部"的喧嚣，进入"内部"的会神凝视思考与宁静的自我存在之中；"漫游"与"沉浸"的状态相互依赖相辅相成，构成层次递进、生动而完整的空间叙事。

"Well-spaced and tortuous" – a profound and touching experience of space and scene, is constantly connected and revealed layer by layer; It guides people to experience the hustle and bustle of the "outside", and enter the self-existence through meditation and peacefulness of the "inside" in a spatial manner; The states of "roaming" and "immersion" are interdependent and constitute a hierarchical, vivid and complete spatial narrative.

视看胜景，最高等级的"沉浸"。一种身心沉浸于人工与自然交互而成的充满时间感、空间感、生命感的深远无尽空间，给人（体验主体）带来的高度诗意状态；以当代方式将人文性赋予自然中的建筑，洞穿时空，浓缩、点化建筑的价值与人性化。

View of the scenery from afar is the highest level of "immersion". A profound and endless space full of a sense of time, of space and of life formed by the coexistence and interaction of the body and mind immersed in the artificial and the naturale, bringing a highly poetic state to people (experience subjects); In a contemporary way, a humaninistic character is given to the building in "nature", penetrating through time and space, condensing and enlightening the value and personification of architecture.

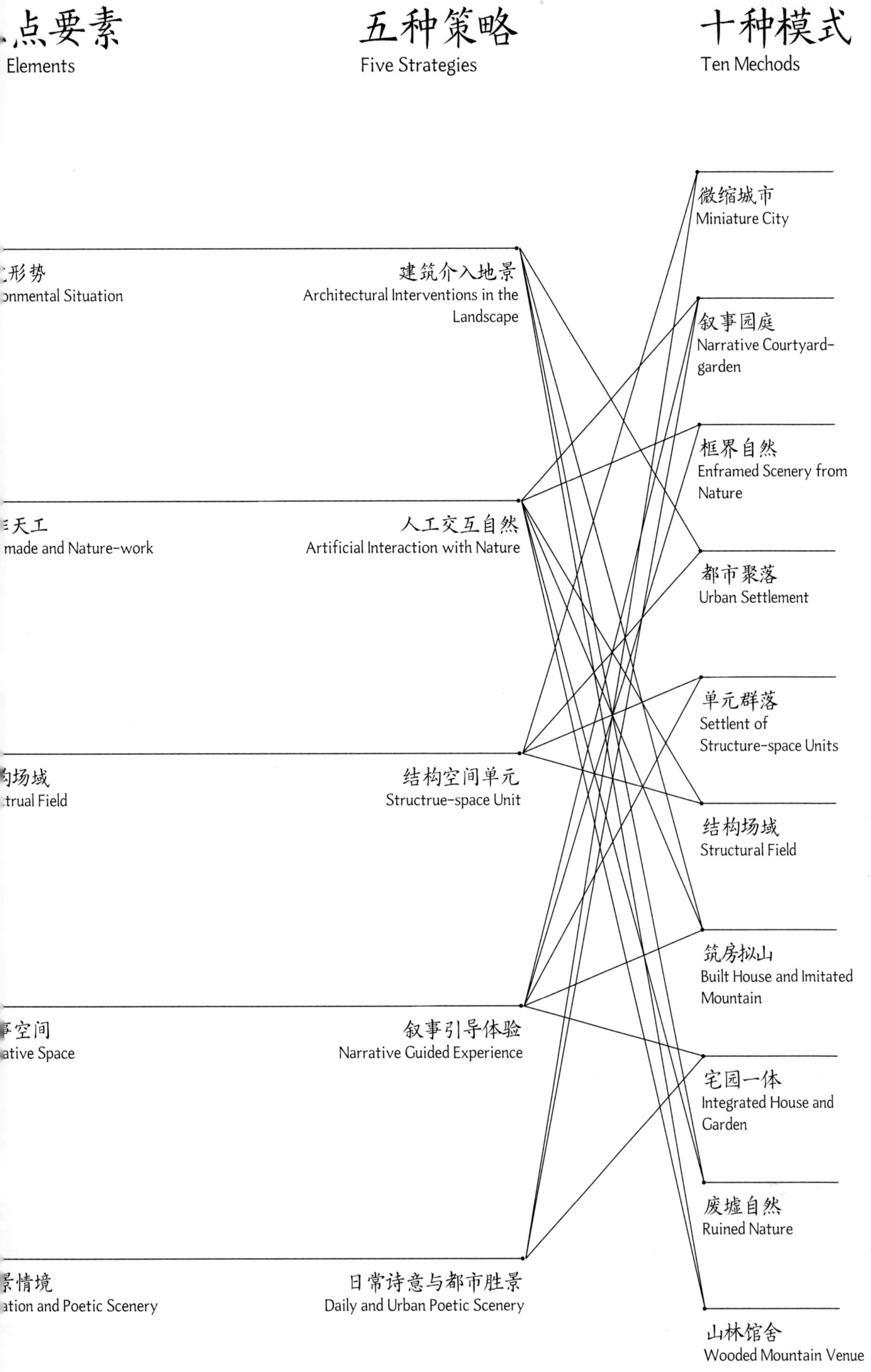

点要素 Elements	五种策略 Five Strategies	十种模式 Ten Mechods
形势 onmental Situation	建筑介入地景 Architectural Interventions in the Landscape	微缩城市 Miniature City
天工 made and Nature-work	人工交互自然 Artificial Interaction with Nature	叙事园庭 Narrative Courtyard-garden
场域 trual Field	结构空间单元 Structrue-space Unit	框界自然 Enframed Scenery from Nature
事空间 ative Space	叙事引导体验 Narrative Guided Experience	都市聚落 Urban Settlement
情境 ation and Poetic Scenery	日常诗意与都市胜景 Daily and Urban Poetic Scenery	单元群落 Settlent of Structure-space Units
		结构场域 Structural Field
		筑房拟山 Built House and Imitated Mountain
		宅园一体 Integrated House and Garden
		废墟自然 Ruined Nature
		山林馆舍 Wooded Mountain Venue

图 2-2

间、胜景情境。

五个策略：建筑介入地景、人工交互自然、结构／空间单元、叙事引导体验、日常诗意与都市胜景[1]。

十种模式：微缩城市、叙事园庭、框界自然、单元群落、都市聚落、结构空间、筑房拟山、宅园一体、废墟自然、山林馆舍[2]。

三、面向现实的理论范式性实践

"胜景几何"建筑理论和"现实理想空间营造范式"（图 2-3）在当下现实环境条件下的应用，呈现出复杂、多

[1] 李兴钢．胜景几何论稿 [M]．杭州：浙江摄影出版社，2020:408.
[2] 李兴钢．胜景几何论稿 [M]．杭州：浙江摄影出版社，2020:126.

图 2-3 胜景城市图（现实理想空间营造范式）
Fig. 2-3 Drawing of Poetic Scenery City （Paradigm for realistic ideal spatial creation）

development of augmentation, and renewal evolution of accumulation.

Five Elements: Environmental Situation, Man-made and Nature-work, Structrual Field, Narrative Space, Situation and Poetic Scenery.

Five Strategies: Architectural Intervention in the Landscape, Artificial Interaction with Nature, Structure-Space Unit, Narrative Guided Experience, Daily and urban Poetic Scenery[1].

Ten Methods: Miniature City, Narrative Courtyard-garden, Enframed Scenery from Nature, Settlement of Structure-space Units, Urban Settlement, Structural Field, Built House and Imitated Mountain, Integrated House and Garden, Ruined Nature, Wooded Mountain Venue[2].

胜景城市

2019.11.22.

图 2-3

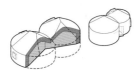

图 2-7

图 2-8

元上都遗址工作站工程，位于内蒙古自治区锡林郭勒元上都遗址明德门之南约1.5 km 的景区入口处。设计以化整为零的分散布局形成了草原上的小群落，一组白色坡顶的圆形和椭圆形小建筑，大小不一、高低错落，围合成对内和对外的两个庭院（图 2-7）。朝向庭院形成沿剖面展开的折线形内界面，暴露内部的混凝土结构；朝向外侧形成连续的弧形界面，外罩白色半透明的 PTFE 膜材，引发蒙古包的联想。建筑以微小、轻盈、临时的自身存在感，表达对宏大、厚重、永固的草原和遗址环境的尊重（图 2-8）。

to the World Heritage Site of Xanadu (Fig. 2-5). The layout, space, and form of the building echo with the mountains, grasslands, and ruins in the external environment, to create an organic interaction and touching dialogue between the artificial, nature, and history (Fig. 2-6).

Visitor's Center for the Site of Xanadu was set up at the entrance of the scenic spot approximately 1.5km south of Mingde Gate. The design is to form a small community on the prairie by shattering the building into parts. A group of small circular and oval buildings with white sloping tops, of different sizes and heights, are enclosed into two courtyards, one for the inside and one for the outside (Fig. 2-7). The tortuous inner interface unfolds along the section towards the courtyard and exposes the internal concrete structure; The continuous arc-shaped interface of the outside is covered with a white translucent PTFE membrane, reminiscent of a Mongolian yurt. The building expresses its respect for the grand, massive, and timeless prairie and ruins environment with its tiny, light, and temporary sense of existence (Fig. 2-8).

图 2-7 元上都遗址工作站工程轴测图
Fig. 2-7 Axonometric drawing of Visitor's Center for the Site of Xanadu
图 2-8 尊重遗址环境的元上都遗址工作站工程
Fig. 2-8 Visitor's Center for the Site of Xanadu which expresses its respect for the prairie

图 2-9

绩溪博物馆工程[1]，位于山水环抱的千年历史的安徽省绩溪县华阳古镇（图2-9）。设计采用"保存与因借"的方式，留树作庭——为尽可能保留用地内的现状树木，整体布局中设置了多个庭院、天井和街巷；折顶拟山——整个建筑覆盖在一个连续的屋面之下，起伏的屋面轮廓和肌理仿佛绩溪周边山形水系，规律性组合布置的三角屋架单元，其坡度源自当地建筑，并适应连续起伏的屋面形态（图2-10）。由古镇民居、周边山水及留存树木共同构成独特的人文机制和自然肌理，与建筑的内外空间、结构、形式交相辉映，使人们体验到有历史时空感的胜景情境（图2-11）。

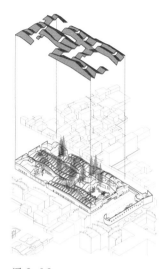

图 2-10

Jixi Museum[1] is located in Huayang Ancient Town, Jixi, a thousand-year-old town surrounded by mountains and rivers (Fig. 2-9). The design adopts the method of "Preserving and Borrowing": Protecting trees to form courtyards-in order to maintain the existing trees on-site as much as possible, several courtyards, patios, and alleys are set up in the masterplan; Folding the roof like a mountain – the whole building is covered under a continuous roof, and the undulating roof outline and texture resemble the mountain-shaped topography around Jixi. The triangular roof truss units are arranged in a regular combination, and the slope is derived from local buildings which is adapted to the continuous undulating roof form (Fig. 2-10). The unique humanistic context with natural texture is formed by ancient dwellings, surrounding landscapes, and preserved trees, which complement each other with the building's interior and exterior space, structure, and form, allowing people to experience the picturesque scene with a sense of historical time and space (Fig. 2-11).

[1] 获得全国工程设计行业一等奖、中国建筑学会建筑创作大奖、中国建筑学会建筑创作奖金奖、北京市优秀工程设计一等奖。
Won the first prize of the National Engineering Design Industry, Architectural Creation Award of the Architectural Society of China, Architectural Creation Gold Award of the Architectural Society of China, and the first prize of Beijing Excellent Engineering Design.

图 2-9 置于古镇中的绩溪博物馆工程
Fig. 2-9 Jixi Museum lying in the ancient town
图 2-10 绩溪博物馆工程轴测图
Fig. 2-10 Axonometric drawing of Jixi Museum
图 2-11 与古镇民居、周边山水、留存树木的融合
Fig. 2-11 Integration with ancient dwellings, surrounding landscapes, and preserved trees

图 2-11

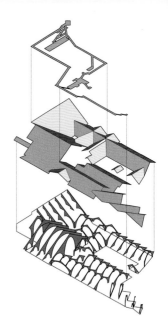

图 2-12

安仁大匠之门文化中心工程，位于千年历史的四川省成都市大邑安仁古镇树人街南端。以安仁古镇独有的公馆建筑类型作为历史参照，塑造了一组自西向东逐渐升高的坡屋顶庭院建筑群，其混凝土片拱结构墙与木屋架结合，形成了既有当代性又有当地性的建构表达（图 2-12）。建筑体量和屋面形态不断转折变换如绵延群山，坡屋顶建筑群的屋顶公共栈道、廊亭可供游人攀览，俯眺古镇老街，并与地面园庭老宅相连接，为人们提供了古镇环境中的立体观游体验，栈道在屋脊处形成供室内展厅通风采光的高侧窗（图 2-13）。

图 2-12 安仁大匠之门文化中心工程轴测图
Fig. 2-12 Axonometric drawing of Anren Culture Center for Great Craftsmen
图 2-13 古镇环境中的立体观游体验
Fig. 2-13 Three-dimensional viewing experience in an ancient town environment

Anren Culture Center for Great Craftsmen is located at the southern end of Shuren Street, Anren Ancient Town, Dayi, Chengdu, Sichuan, with a history of thousands of years. Taking the unique mansion building of Anren Ancient Town as a historical reference, a group of courtyard buildings with sloping roofs is formed, rising from west to east. Its concrete sheet arch structure wall and wooden roof truss are combined to form a construction expression that is both contemporay and reginal (Fig. 2-12). The building volume and roof form are continuously varying, like rolling hills. Tourists can climb up the roof public plank paths and the gallery pavilions of the sloping roof complex for an overlook of the old streets of the ancient town. The roof is also connected to the existing courtyard on the ground, providing people with a three-dimensional viewing experience in an ancient town environment. The plank paths also feature high side windows at the ridge for the ventilation and lighting of the exhibition hall (Fig. 2-13).

图 2-13

第三章

"生态工程原理引导意匠创造"的工程建筑学设计方法

Chapter III

Engineering-Integrated Architectural Design
Method of
"Ecological Engineering Principle Guides
ingenious invention"

延庆赛区俯瞰：山林场馆，
生态冬奥
Overlook of Yanqing Zone:
Mountain Forest Venues /
Ecological Winter Olympics

"生态工程原理引导意匠创造"的工程建筑学设计方法，包括"基于空间耦合的多维空间几何构型"和"基于生态诱导的大型场馆设施与自然环境协调共生"等设计方法，以"生态"应对"失态"，实现具有容纳性的公共空间和生态性的山林场馆，解决了大型建筑场馆对城市生态的排斥和对自然生态的破坏问题。

一、"基于空间耦合的多维空间几何构型"设计方法

　　针对大型复杂建筑在城市"空间生态"环境的孤岛效应问题，提出了"基于空间耦合的多维空间几何构型"设

"Ecological engineering principle guides ingenious invention" contains "multidimensional spatial geometry based on spatial coupling" and "Coexistence of large venues and natural environment based on ecological induction" and other design methods, coping with the "Discordant" "Unsustainable" in a "Concordant" way, realizing accommodating public spaces and ecological mountain forest venues, to avoid the rejection of urban ecology or damage to the natural ecology by large building venues.

I. Design method of "Multidimensional spatial geometry based on spatial coupling"

Aiming to solve the problem of the island effect of large and complex buildings in the urban "spatial ecology" environment, the design method of "Multidimensional spatial geometry based on spatial coupling" was proposed, make the main public buildings have a permeable effect on the urban spatial environment in which they are located, and form a positive urban "Spatial Ecology".

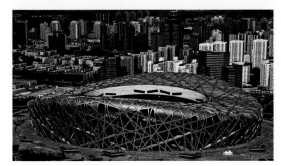

图 3-1

计方法，使重要公共建筑对于其所处城市空间环境产生渗透性效应，形成良好的城市"空间生态"。

北京 2008 年奥运会主体育场暨国家体育场（"鸟巢"）[1]位于向北延伸至奥林匹克公园的古老北京中轴线东侧，是世界瞩目的历史性重大工程之一（图 3-1）。该项目工程设计中，在国内建筑行业首创设立了三维设计专业并领先运用了高精度、参数化 BIM 技术，实现建筑复杂几何构型的秩序化、空间化以及精确设计、定位与实施，并一体化地解决了与之密切相关的膜结构、立面楼梯、体育场微气候、屋顶雨水组织排放系统、消防性能化设计及疏散模拟等一

[1] 获得 RIBA 斯特林建筑大奖、IOC/IAKS 联合金奖、全国优秀工程设计金奖。
Won RIBA Stirling Architecture Award, Gold Medal of IOC IAKS Award, Gold Award of National Excellent Engineering Design.

图 3-1 北京 2008 年奥运会主体育场暨国家体育场"鸟巢"
Fig. 3-1 Beijing 2008 Olympic Games Main Stadium "Bird's Nest"

Beijing 2008 Olympic Games Main Stadium, known as the National Stadium ("Bird's Nest")[1], located on the east side of the ancient Beijing central axis extending northward to the Olympic Park, is a world-famous historic project (Fig. 3-1). In the engineering design of this project, a 3D design methodology has been established for the first time in the domestic construction industry. Meanwhile, the advanced high-precision parametric BIM technology has been used to realize the organization and spatialization of complex geometric structures of buildings and precise design, positioning, and implementation. At the same time, a series of key and closely related technical issues have been solved in an integrated manner, such as membrane structure, facade stairs, stadium microclimate, roof rainwater organization and drainage system, fire performance-based design, and evacuation simulation. Many of these achievements have reached leading international and domestic levels (Fig. 3-2). In the surrounding and even larger urban "spatial ecology" environment, the stadium is like an urban theater or an urban living room (Fig. 3-3),

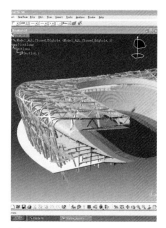

图 3-2

图 3-3

系列关键技术难题，多项成果达到国际、国内领先水平（图3-2）。在周边乃至更大范围的城市"空间生态"环境中，如同城市剧场和城市客厅（图3-3），实现了具有渗透性（吸引）和包容性（友好）特征的超级城市公共空间，并塑造出了有中国文化内涵的当代北京新标志性城市景观。

　　该设计方法还被应用于中国驻新西兰大使馆（惠灵顿）、北京大兴区文化中心、商丘博物馆、中国建筑设计研究院创新科研示范中心等多项城市空间环境中的重要公共建筑中，大大地改善了重要空间生态环境中的公共性和文化性效应。

图 3-2 运用于"鸟巢"的参数化 BIM 技术
Fig. 3-2　Parametric BIM technology using in "Bird's Nest"
图 3-3 作为城市客厅的"鸟巢"
Fig. 3-3　"Bird's Nest" as urban living room

which realizes a super urban public space with permeable (attractive) and inclusive (friendly) characteristics and creates a new iconic urban landscape of contemporary Beijing with Chinese cultural connotations.

The design method has also been applied to many important public buildings in the urban context, such as the Embassy of the People's Republic of China in New Zealand (Wellington), the Cultural Center of Beijing Daxing District, Shangqiu Museum, China Architectural Design and Research Institute Innovation and Research Center, thereby significantly improving the public and cultural benefits in the ecological environment of the main space.

图 3-4

二、"基于生态诱导的大型场馆设施与自然环境协调共生"设计方法

针对大型建筑场馆对于山林"自然生态"环境的破坏效应问题，建立了"基于生态诱导的大型场馆设施与自然环境协调共生"设计方法，使场馆设施与山林环境相得益彰，保持和提升良好的自然生态环境效应。

北京 2022 年冬奥会与冬残奥会延庆赛区[1]，位于燕山山脉军都山以南的小海坨南麓，是北京冬奥会最具体育场地生态挑战性的赛区（图 3-4）。该赛区规划和主要场馆工程设计中（图 3-5），在工程建设行业首创设立了可持续设

[1]"十三五"国家重点研发项目"科技冬奥"重点专项（2018YFF0300300）"复杂山地条件下冬奥雪上场馆设计建造运维关键技术"。
"Thirteenth Five-Year Plan" national key research and development project, "Science and Technology of Winter Olympics" key project (2018YFF0300300), "Key technologies for design, construction, operation, and maintenance of Winter Olympics stadiums under complex mountain conditions".

图 3-4 北京 2022 年冬奥会延庆赛区鸟瞰
Fig. 3-4 Overlook Yanqing Zone of Beijing 2022 Winter Olympic Games
图 3-5 弱介入可逆式装配化高山架空平台技术
Fig. 3-5 Minor intervening reversible assembled alpine aerial platform technology

II. Design method of "Coexistence of the facilities of large venue and natural environment based on ecological induction"

Aiming at the destructive effect of large building venues on mountains and forests' "natural ecological" environment, the design method of "Coexistence of the facilities of large venue and natural environment based on ecological induction" has been established, make the venue facilities complement the mountain and forest environment, and maintain and enhance the enjoyable natural ecological environment effect.

Beijing 2022 Winter Olympic & Paralympic Games Yanqing Zone[1], located at the southern foot of Xiaohaituo south of Jundu Mountain in the Yanshan Mountains, is the most ecologically challenging competition district of the Beijing Winter Olympics (Fig. 3-4). In the planning of the competition area and the engineering design of the main venues (Fig. 3-5), the sustainable

图 3-5

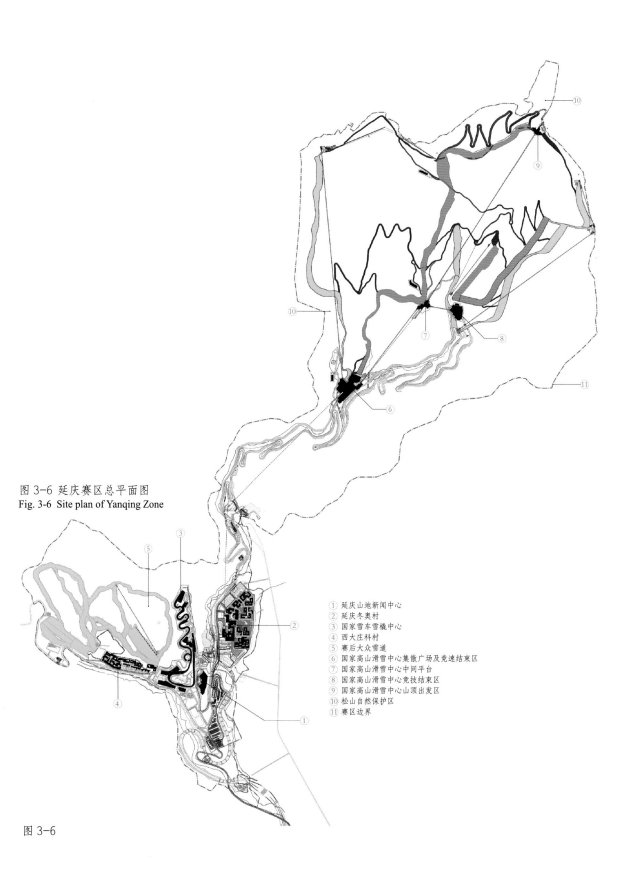

图 3-6 延庆赛区总平面图
Fig. 3-6 Site plan of Yanqing Zone

① 延庆山地新闻中心
② 延庆冬奥村
③ 国家雪车雪橇中心
④ 西大庄科村
⑤ 赛后大众雪道
⑥ 国家高山滑雪中心集散广场及竞速结束区
⑦ 国家高山滑雪中心中间平台
⑧ 国家高山滑雪中心竞技结束区
⑨ 国家高山滑雪中心山顶出发区
⑩ 松山自然保护区
⑪ 赛区边界

图 3-6

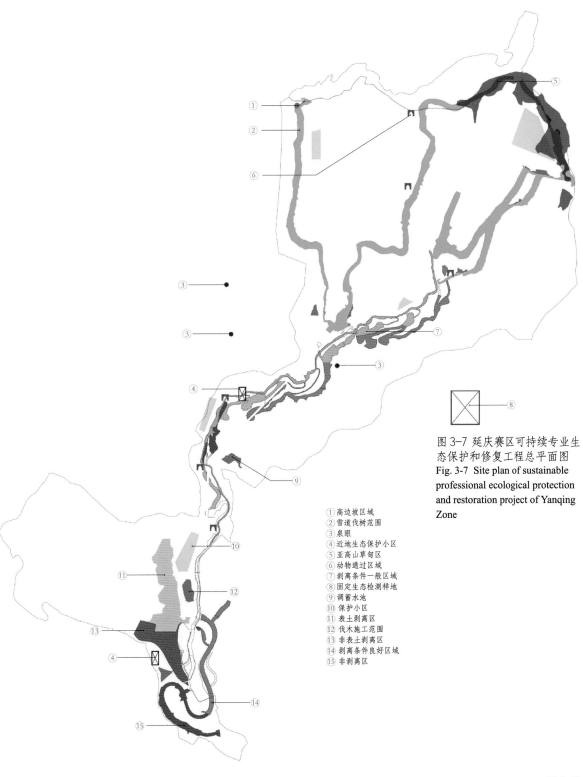

① 高边坡区域
② 雪道伐树范围
③ 泉眼
④ 近地生态保护小区
⑤ 亚高山草甸区
⑥ 动物通过区域
⑦ 剥离条件一般区域
⑧ 固定生态检测样地
⑨ 调蓄水池
⑩ 保护小区
⑪ 表土剥离区
⑫ 伐木施工范围
⑬ 非表土剥离区
⑭ 剥离条件良好区域
⑮ 非剥离区

图 3-7 延庆赛区可持续专业生态保护和修复工程总平面图
Fig. 3-7 Site plan of sustainable professional ecological protection and restoration project of Yanqing Zone

图 3-7

计专业，提出并实现了"体育与生态共生"的创新技术体系（图3-6），研发了复杂山地与赛道场馆高拟合度技术、弱介入可逆式装配化高山架空平台等低地形扰动技术（图3-7）、山地装配化结构及石笼墙等自然建造技术、超大规模木结构和木瓦屋面系统固碳储存先导技术、成套动植物与生态保护及修复技术等关键技术，填补了我国冬奥雪上相关工程建设领域空白，实现场地、场馆和基础设施与自然环境及生态系统的适应和协调，打造了地形复杂、地质脆弱、气候严苛、生态敏感、场馆集约等建设条件下的生态冬奥典范工程，提出了规划—设计—建造—运维全过程冬奥遗

design methodology of the engineering construction industry was established for the first time, and the innovative technology system of "Sports and Ecology Coexistence" was proposed and realized (Fig. 3-6). The high-fit technology of complex mountain and track venues and the low-terrain disturbance technology such as minor intervening reversible assembled alpine aerial platform was developed (Fig. 3-7). The natural construction technology such as prefabricated mountain structures and gabion walls, the pioneering technology for carbon sequestration and storage in ultra-large-scale timber structures and shingle roofing systems, and the complete set of animal, plant, and ecological protection and restoration technology are developed. It has realized the adaptation and coordination of fields, venues, and infrastructure with the natural environment and ecosystem. Meanwhile, it has created a prototype for the ecological Winter Olympics under the conditions such as complex terrain, fragile geology, harsh climate, sensitive ecology, intensive

图 3-8

产长效利用技术体系，形成了国际领先的创新工程技术成果，在最具挑战性的延庆赛区"自然生态"环境中，实现了具有生态性（友好）和文化性（吸引）特征的山林场馆（图3-8）。研发了冬奥标准高难度高复杂度场馆设计、建造和运行的技术体系，高质量满足冬奥赛事需求，建成竞赛场馆被国际单项体育组织认证和评价为"世界领先"。并创作出体现"山林场馆、生态冬奥"理念的当代中国新文化景观。

该设计方法还被应用于北京怀柔水长城书院，崇礼太子城冬奥展示中心及雪花小镇，延庆造雪输水一、二级泵站等自然空间环境中的重要公共建筑，特别是应用于复杂

图 3-8 山林场馆
Fig. 3-8 Mountain Forest Venues

venues, and in the meantime, proposed a long-term utilization technology for the Winter Olympics heritage in the whole process of planning, design, construction, operation, and maintenance. The system has formed the world's leading innovative engineering and technological achievements, realizing the ecological (friendly) and cultural (attractive) mountain forest venues in the most challenging "natural ecology" environment of the Yanqing competition area (Fig. 3-8). It has developed a technical system for the design, construction, and operation of the high-difficulty and high-complexity venues to meet the needs of the high-quality events under Winter Olympics standards. Eventually, the completed venues were certified and evaluated as "world-leading" by International Sports Federations. It has created a contemporary Chinese new cultural landscape that reflects the concept of "Mountain Forest Venues, Ecological Winter Olympics".

The design method has also been applied to important public buildings in the natural environment, such as Beijing Huairou Water Great Wall

山地环境建设及其生态保护与修复工程领域，大大地改善了重要自然生态环境中的生态和人文性效应。

三、基于"双碳"目标的"自能利用、节流开源"设计方法

针对高质量发展背景下大型复杂建筑高能耗、高排放等问题，创立了基于"双碳"目标的"自能利用、节流开源"设计方法和技术，使大型公共建筑降碳、环保，成为人类生存及生活空间环境的有机、友好构成内容。

天津大学新校区综合体育馆工程[1]，位于天津中心城区和滨海新区之间快速建设的海河教育园区天津大学新校

[1] 获得 ArchDaily 年度全球建筑大奖、全国优秀工程勘察设计行业奖一等奖、中国建筑学会建筑创作大奖、中国建筑设计奖（建筑创作）金奖。Won Annual Global Architecture Award of ArchDaily, the first prize of the National Excellent Engineering Survey and Design Industry Award, Architectural Creation Award of the Architectural Society of China, and Gold Award of the China Architectural Design Award (Architectural Creation).

Academy, Snowflake Town of Prince City in Chongli, Yanqing Snowmaking and Water Conveying Primary and Secondary Pumping Stations. It has significantly improved the ecological and cultural effects of critical natural ecological environments, especially for constructions in a complex mountain environment and ecological protection and restoration projects.

III. Design method of "using self-generated energy, reducing consumption and increasing gains" to realize the "double carbon" target.

Under the background of high-quality development, considering the issues of high energy consumption and high emissions of large and complex buildings, the design method and technology of "Using self-generated energy, Reducing consumption and increasing gains consumption" aims to realize the "double carbon" target was created to facilitates large-scale public buildings carbon reduction and environmental protection, and become the organic and friendly components of human survival and living space environment.

The Gymnasium of the New Campus of Tianjin University[1] is located at

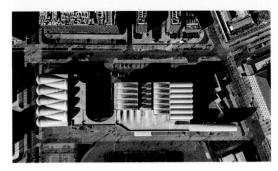

图 3-9

区场地。在极为紧张的用地内，设计采用了一种紧凑高效的空间组合布局方式，依照各类运动场馆空间的平面尺寸、净高及使用工艺，由多组直纹曲面钢筋混凝土薄壳单元构成了空间叠置、功能集约的"立体结构聚落"，营造出体育馆多样而动人的运动空间和建筑自身的存在感、场所感（图3-9），并将自然通风（自动翻转开闭地面活门，自然拔风空腔夹壁墙）、自然采光（大面和连续可开启高侧窗）、自然排水（横纵向排水、汇水天沟、雨水收集）的绿色建筑系统与建筑结构、空间、形态设计巧妙结合，减少了土地占用、能源消耗，并降低了日常维护费用（图3-10）。

the northern end of the new campus of Tianjin University at the Haihe Education Park in the middle stretch of the Haihe River, Tianjin City. The design adopts a compact and efficient space organization in the highly tight land. According to the plane size, clear height, and functional techniques of various sports venues, multiple groups of ruled-surface concrete shells form "three-dimensional structure clusters" with overlapping space and intensive functions, creating a diverse and touching sports space in the gymnasium and a sense of the existence of the building itself (Fig. 3-9). In this way, the green building system of natural ventilation (ground valve with automatically flipping operability, clamping wall with passive air suction), natural lighting (large continuous operable high-side windows), and natural drainage (horizontal and vertical drainage, catchment gutter, rainwater collection) are dexterously combined with the design of building structure, space, and form to reduce land occupation, energy consumption, and maintenance cost (Fig. 3-10).

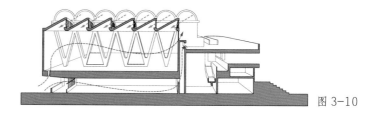

图 3-10

延庆山地新闻中心工程，是位于北京2022年冬奥会延庆赛区内的非竞赛场馆之一（图3-11）。针对狭促紧张的场地条件和环境景观需求，采用了大空间建筑的覆土设计模式，80%的建筑体量掩藏于山体地貌之下，并与周边山体自然衔接，上部形成一个由外露平台、阵列天窗和步道台阶、绿植铺地等构成的山顶公园。其"节流"措施包括种植屋面和围护结构优化等被动式技术降低建筑运行能耗；其"开源"措施包括结合屋顶天窗及平台的光伏一体化系

图 3-10 自然通风、自然采光与建筑设计的结合
Fig. 3-10 Natural ventilation, Natural lighting combined with the design of building
图 3-11 延庆山地新闻中心工程航拍
Fig. 3-11 Aerial image of the Yanqing Mountain News Center

The Yanqing Mountain News Center is one of the non-competition venues located in the Yanqing Zone of the Beijing 2022 Winter Olympics (Fig. 3-11). In response to the narrow and tense site conditions and landscape requirements, the design adopts the earth shelter mode of large-space buildings: 80% of the building volume is hidden under the mountain topography and naturally connected with the surrounding mountains. The upper part forms a hilltop park consisting of external platforms, arrayed skylights, stepped pathways, and vegetated pavement. Its "Reducing consumption" measures include passive technologies, such as planted roofs and enclosed structure optimization, to reduce the energy consumption of building operations. Meanwhile, its "increase gains" measures include integrated photovoltaic systems combined with roof skylights and platforms to generate electric energy to supply its consumption, realizing a near-zero carbon emission building as a demonstration project (Fig. 3-12).

图 3-11

统等产出电能供应自身空间使用，实现了近零碳建筑示范
工程（图3-12）。

　　该设计方法还被应用于瑞典MAX-labⅣ同步辐射国
家实验室项目（隆德）、湖北郧阳博物馆、杭州云城双铁上
盖区域TOD超级城市综合体、深圳留仙洞万科云城北绿廊
03-05地块中区项目、内蒙古通辽美术馆暨蒙古族服饰博
物馆等多项公共建筑工程，探索了节能与产能高度结合的
低碳建筑发展新方向。

The design method has also been applied to the Swedish MAX-lab Ⅳ Synchrotron Radiation National Laboratory Project (Lund), the Yunyang Museum in Hubei, the TOD super city complex above the double railway in Hangzhou Yuncheng, the central area project of Plot 03-05 on the North Green Corridor of Vanke Cloud City in Liuxiandong, Shenzhen, Tongliao Art Gallery and Mongolian Costume Museum, etc. These explored a new direction for the development of low-carbon buildings with high integration of energy-saving and production capacity.

图 3-12 "开源" "节流" 措施
Fig. 3-12 The measures of "increase gains" and "reduce consumption"

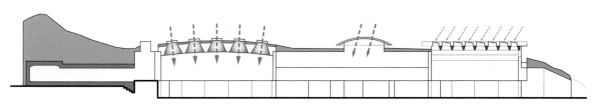

图 3-12

第四章

"技术应用场景引导审美创作"的工程建筑学设计方法

Chapter IV

Engineering-Integrated Architectural Design
Method of
"Technical Application Scenario Guides
Aesthetic Creation"

国家雪车雪橇中心工程鸟瞰
Overlook of the National Sliding
Center

"技术应用场景引导审美创作"的工程建筑学设计方法，包括"数字智慧技术融合场地设计和赛道生成及人体运动""结构及建造技术引导空间生成和形式意匠""功能工艺技术融合城市空间和景观创作"及"环境调控技术引导建筑空间和景观营造"等设计方法，以"有序"应对"失序"，将建筑的空间、形式、景观创造与智能数字、结构建造、功能工艺、环境调控等技术创新互成，形成了技术性与艺术性高度融合的建筑创新路径，解决了由各种复杂因素产生的技术环境失序问题。

The E.I.A. design method of "Technical application scenario guides aesthetic creation" contains "Digital intelligent technology combines site design, track generation, and body movement" "Structure and construction technology guide space generation and form design" "Integrating urban space and landscape creation with functional and process technology" "Environmental control technology guides architectural space and landscape creation" and other design methods. These respond to the "Disorder" with an "Orderly" environment, combining architectural space, form, landscape creation with technological innovations, such as intelligent digital technology, structural construction, functional technology, environmental controls, forming an architectural innovation method with high integration of technology and artistry. It solves the problems of the disordered technical environment caused by various complex factors.

图 4-1

一、"数字智慧技术融合场地设计和赛道生成及人体运动"的设计方法

　　针对复杂地形及超长、超快、超重等特殊体育运动模式带来的超复杂赛场及高难度竞技训练问题，建立了"数字智慧技术融合场地设计和赛道生成及人体运动"的设计方法，使场馆和赛道以高拟合度方式匹配所处复杂地形完成设计建造并最大限度减少对环境干扰，使竞技运动依托赛道形成可精确计量和分析的数字化图像数据用于辅助训练。

　　国家雪车雪橇中心工程[1]，是位于北京2022年冬奥会延庆赛区内的竞赛场馆之一，赛道长度1975米，运行速度

I. Design methods of "digital intelligent technology integrated site design, track generation, and body movement"

In response to the ultra-complex arena and difficult competition training issues caused by complex terrain and unique sport modes such as ultra-long, ultra-fast, or overweight conditions, the design method of "Digital intelligent technology integrated site design, track generation, and body movement" was established to make the venue and track match the complex terrain in a high fit manner to complete the design and construction, and minimize interference with the environment, so as to generate digital track image data that can be accurately measured and analyzed for the auxiliary training of competitive sports.

The National Sliding Center[1] is one of the competition venues in the Yanqing Zone of the Beijing 2022 Winter Olympics. The track length is 1,975 meters, the maximum speed is 135 km/h, and the maximum acceleration is 4.9g (Fig. 4-1). A complete set of digital technical methods and intelligent platforms

[1] "十三五"国家重点研发项目"科技冬奥"重点专项（SQ2021YFF030287）"智能雪车雪橇赛道与竞技训练关键技术研究"。
"Thirteenth Five-Year" national key research and development project, "Science and Technology of Winter Olympics" key project (SQ2021YFF030287) "Key technologies for design, construction, operation and maintenance of Winter Olympics stadiums under complex mountain conditions".

图 4-1　国家雪车雪橇中心工程鸟瞰
Fig. 4-1 Overlook of National Sliding Center

图 4-2

最高135千米/小时，最大重力加速度49g（图4-1）。该工程应用了成套数字化技术手段和智慧平台，构建了"复杂地形建模方法 - 新型数字化融合设计方法 - 特殊构造数字化模拟测算方法 - 与国家队科学训练结合的赛道数字孪生方法"等多流程阶段方法，研发了智能雪车雪橇赛道设计与竞技训练关键技术，包括赛道数字化选线成型、复杂场地与场馆BIM融合及GIS一体化协同（图4-2）、高精度三维曲面超长薄壳赛道数字化生成及一体化成型（图4-3）、车橇赛道"地形气候保护系统"、基于高精度智能感知和数字孪生模型的车橇滑行竞技训练即时化、可视化

图 4-3

have been applied, and multi-process staged methods such as "complex terrain modeling method - new digital fusion design method - special structure digital simulated calculation method - track digital twinaided national team training" has been constructed. It has developed leading key technologies for innovative bobsleigh and sled track designs and competitive training, including digital track route selection and formation, BIM integration of complex fields and venues and GIS integration and collaboration (Fig. 4-2), digital generation and integrated molding of high-precision three-dimensional curved ultra-long thin-shell track (Fig. 4-3), "Terrain Weather Protection System" for bobsleigh track, real-time visualization and correction of skid skiing competitive training based on high-precision intelligent perception and digital twin model technology, etc. These methods have guided and supported the design and construction of venues and tracks with innovative digital technologies and the trainning of athletes for the Olympics.

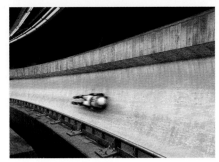

图 4-4

纠偏等国际前沿技术，以数字智慧技术引导支撑场馆和赛道设计建造及运动员奥运备战。国家雪车雪橇中心被各国顶尖运动员群体评价为"全球顶尖的滑行中心赛道和场馆"（图 4-4）。

　　该设计方法还被应用于厦门音乐中心、重庆两江体艺中心等复杂场地大型公共建筑和顶级滑行中心场馆的设计建造及全季节大众滑行运动体验领域。

二、"结构及建造技术引导空间生成和形式意匠"的设计方法

　　针对大空间、大跨度建 / 构筑物中建 / 构筑形式与空

The National Sliding Center has been rated as "the world's top sliding track and venue" by top athletes worldwide (Fig. 4-4).

The design method has also been applied to the design and construction of large-scale public buildings in complex sites such as Xiamen Music Center, Liangjiang Sports and Arts Center of Chongqing, and the design and construction of the top skating center and venues, as well as all-season public skating sports facilities.

II. Design methods of "structure and construction technology guide spatial generation and form design"

In response to the problem of the discrete and fragmented design of the construction form and space utilization, structural construction, and other elements in large-space and large-span buildings/structures, a design method of "Structure and construction technology guide spatial generation and form

图 4-2 复杂场地与场馆 BIM 融合及 GIS 一体化协同
Fig. 4-2 BIM integration of complex fields and venues and GIS integration and collaboration
图 4-3 高精度三维曲面超长薄壳赛道数字化生成及一体化成型
Fig. 4-3 Digital generation and integrated molding of high-precision three-dimensional curved ultra-long thin-shell track
图 4-4 运动员在赛道上滑行
Fig. 4-4 Athlete sliding on the track

图 4-5

间利用、结构建造等要素相互离散、割裂设计的问题，建立了"结构及建造技术引导空间生成和形式意匠"的设计方法，建筑的空间、形式生成依托结构建造技术形成建筑创意特色，实现建筑设计与结构建造的一体化。

海南国际会展中心工程[1]，位于海南省海口市西部新的城市组团北部城市南北向景观轴线的尽端（图 4-5）。在整体布局和空间构成上采用一体化的设计手法，将展览中心和会议中心"聚零为整"，整合处理为一个巨大的完形体量。将大跨度屋面分解为由露明密格钢管所构成的多组连续双向正余弦曲面极薄拱壳单元，形成了最适宜跨度、最

[1] 获得全国优秀工程勘察设计行业奖一等奖、中国建筑学会建筑创作大奖、中国土木工程詹天佑奖、中国建筑设计奖（建筑创作）金奖。Won the first prize of the National Excellent Engineering Survey and Design Industry Award, Architectural Creation Award of the Architectural Society of China, China Civil Engineering Zhan Tianyou Award, and Gold Award of the China Architectural Design Award (Architectural Creation).

图 4-5 海南国际会展中心工程鸟瞰
Fig. 4-5 Aerial view of Hainan International Convention & Exhibition Center

design" was established. The space and form generation of the building relies on the structural construction technology to form creative architectural features, to realize the integration of architectural design and structural construction.

The Hainan International Convention & Exhibition Center[1] is located at the end of the north-south landscape axis of the new city group west of Haikou City (Fig. 4-5). In terms of the overall layout and spatial composition, the design method of integration is adopted: the exhibition center and the conference center are "joined into one" and integrated into a vast united volume. The large-span roof is decomposed into multiple groups of ultra-thin arch shell units composed of naked steel pipes with continuous two-way sine and cosine curved surface, forming an integrated convention and exhibition space with the most efficient span, the most economical clear height, and the most suitable architectural form for the seaside temperature. At the same

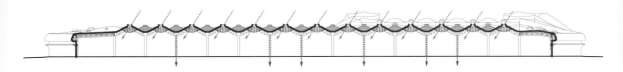

经济净高的会议展览空间以及与海滨场地气质最为契合的建筑形态，并利用拱顶天窗提供自然采光、拱壳底部雨水通过隐藏于钢柱内的水管收集回用，体现出建筑空间、结构、形式、功能等的高度一体化特征（图 4-6）。

　　玉环博物馆和图书馆工程，位于浙江省台州市的海岛县——玉环的填海新城开发区。在近乎"荒芜"的环境中，设计移植了当地坎门渔港的空间形态和意象，博物馆和图书馆则呈现为围绕"渔港"以单元群落方式组构而成的现代"渔村聚落"。两组建筑群置放在巨大的石砌基座之上，由反曲面混凝土悬索结构和大跨度鱼腹梁结构作为基本结

图 4-6 建筑空间、结构、形式、功能的高度一体化
Fig. 4-6 The highly integrated characteristics of architectural space, structure, form, and function

time, vaulted skylights are used to provide natural lighting, and the rainwater is collected and reused through water pipes hidden in steel columns at the bottom of the arch shell. These reflect highly integrated characteristics of architectural space, structure, form, and function (Fig. 4-6).

The Yuhuan Museum & Library is located in a reclamation new town development Zone in the seaside county of Yuhuan, Taizhang City, Zhejiang Province. In an almost "barren" environment, the design transplants the space configuration and images from the local Kanmen fishing port. The museum and library are presented as modern "fishing village settlements" organized around the "fishing port" in the form of unit communities. The two clusters of buildings are placed on huge stone bases, and the reverse-curved concrete suspension cable structure and the large-span fish-beam structure are used as the basic structures and function, space and form units. The structural and spatial units are repeatedly combined, mutated, connected and enclosed

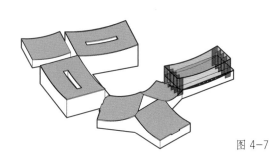

图 4-7

图 4-8

构与功能、空间、形式单元，在水平和垂直两个方向反复组合、变异、连接与围合，形成了独具特色的室内无柱空间和室外群组空间（图 4-7）。两组建筑的内庭相对遥望，并由长长的景观水面连通起来，相互形成彼此的对景（图 4-8）。

上海临港星空之境公园日月桥和无限桥工程，位于上海自贸区临港新片区星空之境公园内。日月桥位于星空之境公园东北部，由两个主副桥互相倚靠相切而成，二者分别服务于车行和人行，并在顶部交汇贯通（图 4-9），弧线性主桥采用了单边支撑的索承桥结构体系，利用钢索顺

in horizontal and vertical directions, forming unique indoor column-free spaces and outdoor cluster spaces (Fig. 4-7). The central courtyards of the two groups of buildings look out in the distance, and are connected by a long water landscape, forming opposite views of each other (Fig. 4-8).

Riyue Bridge and Infinite Bridge are located in the Starry Sky Park, Lingang Free Trade Zone, Shanghai. The Riyue Bridge, located in the northeast of the park, is formed by one main bridge and one auxiliary bridge leaning against each other, serving vehicles and pedestrians respectively and meeting at the top (Fig. 4-9). The arc-shaped main bridge adopts a cable-supported structure system supported on one side, and keeps a balance of forces between the upward rotating component force along the radial direction of the bridge deck and the gravity of the bridge's ownweight to maintain a stable state. Meanwhile, the adjacent auxiliary arched bridge supports the inclined columns across the river, ingeniously realizing a minimal intervention

图 4-9

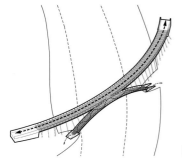

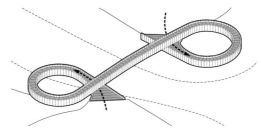

图 4-10　　　　　　　　　　　　　　　　　　　　　　　图 4-11

弧形桥面的径向产生向上转动的分力与桥面自重来维持受力平衡的稳定状态，并利用相靠的拱形副桥支撑跨河段的斜柱，巧妙实现了大跨度桥体对河道和生态的最小化介入（图 4-10）；无限桥位于星空之境公园的中心位置，行人身处桥上可以放眼远眺整个公园。桥身形态借鉴了数学中的无限符号"∞"，隐喻和诠释宇宙星空的"无限"主题，分为两个圆弧上坡段和一个跨河直桥段，采用钢与混凝土组合结构创新技术，并结合场地自然巧妙连接了河道两岸，如轻盈的丝带一般飘落水面，体现了科技感和未来感（图 4-11）。

图 4-10 日月桥工程轴测图
Fig. 4-10 Axonometric drawing of Riyue Bridge
图 4-11 无限桥工程轴测图
Fig. 4-11 Axonometric drawing of Infinite Bridge

of the long-span bridge body to the river-bed channel and ecology (Fig. 4-10). The Infinity Bridge is located in the center of the park, where pedestrians can overlook the entire park from the bridge. The shape of the bridge body draws on the mathematical symbol of infinity "∞", creating a metaphor and interpretation of the "infinity" theme of the cosmic starry sky. The bridge is divided into two parts arcs uphill and one straight part across the river with an innovative steel and concrete composite structure. The bridge connects the two sides of the river banks naturally and slightly, like a light ribbon floating on water, illustrating a sense of technology and the future (Fig. 4-11).

The design method has also been applied to many large-space and large-span public building projects, such as the renovation of Chengdu Sports Center, Shanxi Luliang Sports Center, and the National Stadiums World Championship Registration Center, leading a creative direction of integrated architectural design and operation efficiency.

该设计方法还被应用于成都体育中心改造、山西吕梁体育中心、国家体育场世锦赛注册中心等多项大空间、大跨度公共建筑工程，引领了建筑一体化设计并融合运行效能的创新方向。

三、"功能工艺技术融合城市空间和景观营造"的设计方法

针对特定功能工艺条件下城市公共空间和标识性特征营造的问题，建立了"功能工艺技术融合空间和景观营造"的设计方法，建筑的空间、形式创作依托功能工艺技术形成建筑创意特色，实现建筑设计与功能工艺的一体化。

III. Design methods of "Integrating urban space and landscape creation with functional and process technology"

In response to the problem of building public space and iconic features under specific functional and technological conditions, the design method of "Integrating urban space and landscape creation with functional and process technology" was established. The space and form creation of buildings rely on functional technology to form architectural creative characteristics, and realize the integration of architectural design and functional technology.

Xihuan Plaza & Xizhimen Transport Hub, located in the northwest corner of the Beijing Xizhimen overpass, was China's first large-scale comprehensive passenger transport hub. It mainly serves urban rail transit combined with

图 4-12

北京西直门交通枢纽暨西环广场工程，位于北京市西直门立交桥西北角，是我国第一座以城市轨道交通换乘为主并与商业综合体相结合的大型综合客运交通枢纽，涉及建筑设计、交通规划、流量分析、高架路桥、轨道交通等多专业领域的密切合作。首次在国内实现了国铁、城铁、地铁、公交等多种交通方式共同参与的"零换乘"运行模式（图 4-12），建筑总体布局同时考虑了城市视线通廊，塔楼建筑的立面和屋顶既有创新又暗示了北京传统的建筑屋顶形式，为北京此区域提供了新的城市轮廓线，并依托交通枢纽成为便捷、集聚、共享的大型公共空间、新兴活

图 4-12 "零换乘"运行模式
Fig. 4-12 The "zero-transfer" operation mode

commercial complexes, involving close cooperation in multi-disciplines such as architectural design, traffic planning, flux analysis, devated infrastructure, rail transit, etc. For the first time in China, the "zero-transfer" operation mode with various transportation modes such as the national railway, suburban railway, subway, and lines of bus had been realized (Fig. 4-12). The overall layout of the building also considers the city's visual corridors. The facade and roof form of the towers are both innovative and suggestive of Beijing's traditional architectural roofs, providing a new urban outline for this area of Beijing. At the same time, with the transportation hub, the building becomes a convenient, agglomerated, and shared large-scale public space, an

图 4-13

图 4-14

力中心和城市标志性工程（图 4-13）。

　北京地铁昌平线西二旗站工程[1]，是北京市地铁昌平线的南起终点站，也是与已建成运行的城铁 13 号线的换乘站（图 4-14）。研究确定的车站形式为四柱三跨框架式高架车站，其中昌平线部分为高架侧式站台（二层），城铁 13 号线部分为地面侧式站台（一层），车站的半透光双矩形组合筒状 PTFE 膜结构源于最佳交通换乘流线和"折纸"原理，最大程度匹配了城市地铁终点站和换乘站人流量大、换乘关系复杂的交通工艺需求（图 4-15），实现了模数化、标准化、预制化的设计与建造，并因其日间向内透光、夜间

[1] 获得 IFAI 杰出成就奖、北京市优秀工程设计一等奖。
Won Outstanding Achievement Award of IFAI, the First Prize of Beijing Excellent Engineering Design.

图 4-13 作为城市标志性工程的北京西直门交通枢纽暨西环广场工程
Fig. 4-13 Xihuan Plaza & Xizhimen Transportation Exchange Hub as a city landmark
图 4-14 北京地铁昌平线西二旗站工程
Fig. 4-14 Xi'erqi Station of Changping Line of Beijing Subway
图 4-15 最大程度匹配交通工艺需求的建筑设计
Fig. 4-15 Architecture design maximizes the matching of complex transfers

emerging vitality center, and a city landmark (Fig. 4-13).

Xi'erqi Station of Changping Line of Beijing Subway[1] is the southern terminus of the Changping Line and a transfer station with the formerly completed Metro Line 13 (Fig. 4-14). The station form as determined by study is a four-column three-span frame-type elevated station, of which the Changping Line part is an elevated side platform (second floor) and the urban railway line 13 part is a ground side platform (first floor). The station's semi-transparent PTFE membrane structure in a double-rectangular combined cylindrical shape was derived from the best traffic transfer circulation and "folding paper" principle. It maximizes the matching of urban subway terminals and transfer stations with large traffic flux and complex transfers (Fig. 4-15). The project realizes modularization, standardization, and prefabrication in the design and construction. With the characteristics of inward daylight transmission during the day and outward artificial light transmission at night, and its

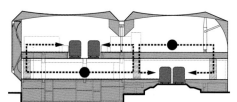

图 4-15

向外透光的特性，以其"弱"和"轻"的建造方式和不动声色的建筑姿态，成为既节约能耗又独具标识性的重要城市公共节点。

北京朝阳区垃圾焚烧发电中心工程，位于北京市朝阳区金盏乡高安屯村生活垃圾综合处理厂区，是一座拥有国际先进设备及工艺的现代大型垃圾焚烧中心。垃圾是人类城市生活的必然产物，随着城市的快速发展，垃圾处理中心作为越来越重要的城市基础设施之一，亟待改变其刻板单一、与污染关联的厂房印象。设计通过对内部工艺和外部围护系统的整合，使建筑体量满足了设备工艺对空间的

"gentle" and "light" construction methods and quiet architectural posture, the building has become an essential urban public space that not only limits energy consumption but also becomes a unique public landmark.

Waste Treatment Center in Chaoyang, located in the domestic waste comprehensive treatment plant in Gao'antun Village, Jinzhan, Beijing, is a modern large-scale waste incineration center with internationally advanced equipment and technology. Waste is an inevitable product of urban life. With the rapid development of urbanization, the waste disposal center, as an increasingly crucial urban infrastructure, urgently needed to change its stereotyped factorylike image, which was always associated with "pollution". The design makes the building volume meet the space requirements of the equipment and technology by integrating the internal treatment process and the external envelope to present the technological process

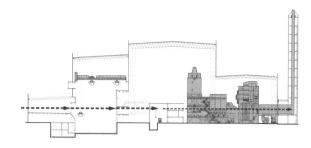

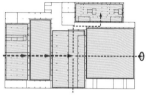

图 4-16

图 4-16 匹配垃圾运卸、存放、焚烧、净化、发电工艺流程的建筑设计
Fig. 4-16 Architecture design matches the process of waste transportation, storage, incineration, purification, and power generation

图 4-17 城市市政公用设施新形象
Fig. 4-17 A new image of public facilities to the public

要求，将垃圾运卸、存放、焚烧、净化、发电的工艺流程及空间特征呈现出来（图 4-16），选用表征工业建筑特征的镀铝锌圆浪形截面波纹钢板作为立面建材并作竖向肌理排布，强调了体量的完整性和工业建筑的尺度感、力量感，向公众传递了健康、可持续的城市市政公用设施新形象（图 4-17）。

该成果被推广应用于北京地铁 4 号线及大兴线地面出入设施等交通、市政基础设施等多类型公共建筑工程，示范了大型公共建筑的功能工艺融合城市公共设施及建筑创新的理性设计方向。

and spatial characteristics of waste transportation, storage, incineration, purification, and power generation (Fig. 4-16). The building uses aluminized zinc corrugated steel plate with a circular wave section that represents the characteristics of industrial buildings as the facade building material, which is arranged as vertical textures. Meanwhile, it also emphasizes the integrity of the volume and the sense of scale and strength of industrial buildings to convey a new image of healthy and sustainable urban facilities to the public (Fig. 4-17).

This achievement has been applied to various public building projects, such as ground access facilities of Line 4 & Daxing Line of Beijing Subway and urban infrastructures, demonstrating the rational design direction of large-scale public buildings' functional technology integrated with urban public facilities and architectural innovations.

图 4-17

四、"环境调控技术引导建筑空间和景观营造"设计方法

针对典型地域气候条件下大型公共建筑的内部空间环境热舒适度保持问题，建立了"环境调控技术引导建筑空间和景观营造"设计方法。建筑的空间、景观营造依托环境调控技术形成建筑创意特色，实现建筑设计与环境调控的一体化。

威海名座大厦工程[1]，位于山东半岛的威海市东部海滨。设定了多组立体"风径"空间，利用当地的主导风向、气压差原理和热压"烟囱"效应（图 4-18），在设计过程中使用了 CFD 计算机模拟技术对风速、温度、湿度等进行

IV. Design method of "Environmental control technology guides architectural space and landscape creation"

Aiming at the problem of maintaining thermal comfort in the interior spatial environment of large public buildings under typical regional climate conditions, the design method of "Environmental control technology guides architectural space and landscape creation" was established. The space and landscape of the building acquire architectural creative characteristics with environmental control technology, and realize an integration of architectural design and environmental control.

The "Hiland · Mingzuo"[1] project in Weihai is located on the east coast of Weihai City, Shandong Peninsula. Multiple sets of three-dimensional "wind path" spaces are set up with the local dominant wind direction, the principle of air pressure difference, and the "chimney" effect of thermal pressure

[1] 获得全国优秀工程勘察设计行业奖二等奖。
Won the Second Prize of the National Excellent Engineering Survey and Design Industry Award.

图 4-18 立体"风径"空间
Fig. 4-18 three-dimensional "wind path" space

图 4-18

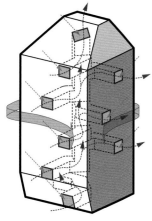

图 4-19

图 4-20

舒适度模拟验证和校核, 并在建成后进行了实际使用效果测试评估, 有效引入了夏季自然风穿越建筑内部以降温除湿, 从而取消了中央空调, 同时回避了冬季风的不利影响, 并形成了多层空中的邻里交往公共空间和面向大海和城市的观景平台 (图 4-19), 使风、光、景观等自然要素和人的活动在建筑中交融共存, 也使建筑以一种开放的姿态和独特的形象存在于城市之中 (图 4-20)。

中国驻西班牙大使馆办公楼改造工程, 位于西班牙首都马德里市中国驻西班牙大使馆院区内。设计以楼梯、廊道和 "景墙" 为焦点元素, 形成了一个自下而上的完整公共空间系

(Fig. 4-18). In the design process, CFD computer simulation technology was used to simulate and verify the comfort level of wind speed, temperature, humidity, etc. After the complete evaluation of the actual operation, the natural summer wind can be effectively introduced to pass through the building to cool down and dehumidify the inside, thus unnecessitating the central air conditioning system and avoiding the adverse effects of winter wind. Meanwhile, several floors of midair public spaces for neighborhood communications and viewing platforms facing the sea and the city are formed (Fig. 4-19). Eventually, natural elements such as wind, light, landscape, and human activities can coexist to make the building stand in the city with an open attitude and a unique image (Fig. 4-20).

The transformation of the Chinese Embassy in Spain office building is located in the campus of the Chinese Embassy in Spain in Madrid. The design focuses on stairs, corridors, and "scenery walls", forming a complete bottom-up public space system that reshapes the interior space. In order to solve the

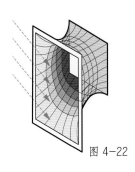

图 4-22

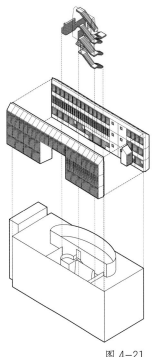

图 4-21

统，重塑了内部空间。为解决现状办公楼东、西晒问题，在原建筑立面外悬挂覆盖了一整套预制 GRC 立体遮阳构件（图 4-21）。其三维几何形态基于结构模数和开窗尺寸，在窗内外两个矩形界面之间自动生成了最简曲面，形成了团扇形的"框景"（图 4-22），为办公空间内工作的人们呈现中国传统"景窗"画意，并向西班牙天才建筑师高迪致敬（图 4-23）。

　　该设计方法还被应用于建川博物馆聚落暨汶川地震纪念馆等多项国内外办公、商业、体育等量大面广的公共建筑工程，充分挖掘和利用地域气候条件，形成了建筑空间、形式创作与景观营造的创新特色。

problem of east and west sun exposure of the current office building, a set of prefabricated GRC three-dimensional sunshade components were hung up and covered the original building facade (Fig. 4-21). Its three-dimensional geometric form is based on the structural modulle and the size of the window openings. The minimal surface is generated between the two rectangular interfaces inside and outside the window, forming a fan-shaped "enframed scene" (Fig. 4-22), representing the Chinese traditional "Scenic Window" for people working in the office space and paying tribute to the genius Spanish architect Gaudí (Fig. 4-23).

This design method has also been applied to other domestic and foreign public, commercial, and sports building projects, such as Jianchuan Mirror Museum & Wenchuan Earthquake Memorial. By fully studying and utilizing the regional climatic conditions, innovative features of architectural space, form innovation, and landscape creation have been formed.

图 4-21　公共空间系统与一套预制 GRC 立体遮阳构件
Fig. 4-21　Public space system and a set of prefabricated GRC three-dimensional sunshade components
图 4-22　单个 GRC 立体遮阳构件
Fig. 4-22　Prefabricated GRC three-dimensional sunshade component
图 4-23　中国驻西班牙大使馆办公楼改造工程
Fig. 4-23　The transformation of the Chinese embassy in Spain office building

图 4-23

第五章

"遗产基因机制引导空间胜景"的工程建筑学设计方法

Chapter V

Engineering-Integrated
Architectural Design Method
of
"Inherited Genetic Mechanism Guides
Poetic Scenery of Space"

北京大院胡同28号改造工程
夜景鸟瞰
Aerial view of the night scene of
reconstruction project of No.28,
Dayuan Hutong, Beijing

"遗产基因机制引导空间胜景"的工程建筑学设计方法，包括历史遗产"新旧相生、长效利用"、旧城疏解改造"分形加密、重建规制"、居住建筑"理想空间、当代重构"和乡村建造"空间记忆、在地建造"等可持续设计方法，以"平衡"应对"失衡"，强调原生自然环境和人工自然环境对建筑设计的引导交互作用，再造人居环境中可体验感知的空间胜景，解决了当代城乡复杂多元的空间环境失衡问题。

一、历史遗产"新旧相生、长效利用"设计方法

　　针对存量城市发展与更新中历史遗产的保护与利用问

Engineering-integrated architectural design method of "Inherited Genetic Mechanism Guides Poetic Scenery of Space" contains the sustainable design methods of realizing the "coexistence of the old and new with long-term utilization" to historic heritage, applying "fractional encryption and morphology reconstruction" to the old urban context, creating an "ideal space and contemporary reconstruction" for residential buildings, and inheriting "spatial memory and local identity" for rural development, etc. It uses the "Balanced" to deal with "Imbalanced" conditions, to guide interactions between the native natural environment and artificial natural environment in architectural designs, recreating the spatial poetic scenery that can be experienced and perceived in the human settlement. As a result, it solved the imbalance of diversified complexity in contemporary urban and rural spatial environments.

I. The design method of realizing "coexistence of the old and new with long-term utilization" for historic heritage

In response to the protection and utilization of historic heritage in the development and renewal of existing cities, sustainable design methods and

图 5-1

题，建立了"新旧相生、长效利用"的可持续设计方法和技术，强调对历史遗产的珍视保护和与新建筑的融合共生，延续历史记忆、复兴城镇活力。

首钢工舍工程[1]，位于北京市石景山区首钢搬迁后遗留的工业园区（图5-1）。采用了"珍视保护，新旧叠加"的设计策略，将被废弃的四座工业建筑遗产作为社会、时空记忆的载体和"人文自然"遗产基因，使被保留利用的机站旧仓与叠加其上的客房新阁相融共生，分别容纳新的使用功能（图5-2）。极具工业特色的构件被戏剧性地暴露在公共空间中，对原建筑进行全面结构检测基础上，确定了"拆除、加固、保留"相结合的结构处理方案；使用粒子喷

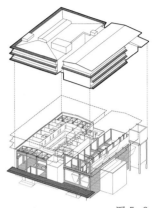

图 5-2

technologies of "coexistence of the old and new with long-term utilization" were established, which emphasizes the cherishings and protection of historic heritage and the integration and symbiosis with new buildings, so as to maintain historical memory and revitalize the town.

The "Silo Pavilion" Holiday Inn Express Beijing Shougang[1] is located in the industrial park, in shijingshan District, Beijing which was preserved after the relocation of Shougang (Fig. 5-1). The design strategy of "cherishing protection, superimposing the old and new" was adopted. Four abandoned industrial heritage buildings were used as the carrier of social, space-time memory, and the inherited gene of "humanity and nature". The preserved and reused old warehouse of the factory and the new guest rooms superimposed on the warehouse are integrated and symbiotic with new functions (Fig. 5-2). The components with industrial characteristics are dramatically exposed in the public space. Based on a comprehensive structural inspection of the original building, a structural treatment plan combining "demolition, reinforcement, and retention" was determined. By using particle spray to

[1] 获得"世界未来城市计划"（IUPA）提名奖、中国建筑设计奖（历史文化保护传承创新类）一等奖、中国建筑设计奖（公共建筑类）一等奖。
Won the "World Future City Plan" (IUPA) nomination award, the first prize of China Architectural Design Award (Historic and Cultural Protection, Inheritance, and Innovation Category), the first prize of China Architectural Design Award (public building category).

图 5-1 首钢工舍工程
Fig. 5-1 The "Silo Pavilion" Holiday Inn Express at Beijing Shougang
图5-2 "珍视保护，新旧叠加"的设计策略
Fig. 5-2 The design strategy of "cherishing protection, superimposing the old and new"

图 5-3

射技术对需保留的涂料外墙进行清洗，在清除污垢的同时，保留了数十年形成的岁月痕迹和历史信息。该项目成为首钢园区乃至北京城市复兴的样板工程。

　　<u>四川廖维公馆改造暨安仁古镇游客中心工程</u>，位于四川省成都市附近安仁古镇的边缘一处现状农田之中，并比邻一个随后建成的大型居住社区（图5-3）。设计最大程度保留了廖维公馆及其庭院绿植，将荒废的老建筑由私人宅邸改造成向公众开放的建筑，沿不断升高、尺度变大的三进原有宅院格局的两端各自扩建，形成了一个高低渐变、新旧融合的坡屋顶整体群落（图5-4），作为社区公共活动

图 5-3 置于古镇中的四川廖维公馆改造暨安仁古镇游客中心工程鸟瞰
Fig. 5-3 Overlook of the renovation of Liao Wei mansion and Anren Ancient Town Visitor Center located in the ancient town

图 5-4 高低渐变、新旧融合的坡屋顶整体群落
Fig. 5-4 An integrated cluster of sloping roofs with a gradual change in height and a fusion of the old and new

clean the painted exterior walls, which need to be preserved, the traces of years and history formed over decades are maintained. It becomes a demonstration project for Shougang Park and Beijing's urban revival.

Renovation of Liao Wei mansion and Anren Ancient Town Visitor Center, located in an existing farmland on the edge of the ancient town of Anren near Chengdu, is adjacent to a large residential community built at a later time (Fig. 5-3). The design preserves the Liao Wei Mansion and its courtyard green plants to the maximum extent. It transforms the abandoned existing buildings from private residences into buildings open to the public. The buildings are expanded along the two ends of the original three-courtyard layout, continuously rising and increasing in size, and form an integrated cluster of sloping roofs with a gradual change in height and a fusion of the old and new (Fig. 5-4). The new mansion's spatial interface as the community's

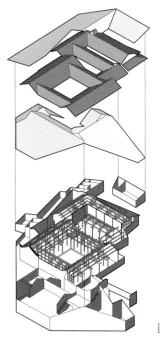

图 5-4

中心的空间界面尺度亲切，而面向城镇空间的游客中心则形成尺度巨大的坡屋顶入口，并与由此延展而生的社区建筑群落围合出街巷、广场，建筑与整个社区新旧融合共生，成为中国西南地区新城镇发展的研究性样本。

　　该设计方法还被应用于北京东城区工人文化宫改造、泉州当代艺术馆旧馆、北京复兴路乙59-1号改造、中共洛阳组诞生地党史纪念馆、中国驻爱沙尼亚大使馆（塔林）等多项公共建筑工程，对历史遗产建筑改造利用并延长其使用生命周期，从而实现了建筑降碳目标，并创作出与旧建筑共生的当代新建筑。

public activity center is friendly in scale. The visitor center facing the urban space forms the entrance with a substantial pitched roof. The extended community buildings enclose streets, alleys, and squares. The buildings and the whole community coexist as an integration of the old and new, becoming a research model for developing new towns in Southwestern China.

The design method has also been applied to the renovation of the Workers' Cultural Palace in Dongcheng District, Beijing, the renovation of Quanzhou Museum of Contemporary Art, the renovation of No. B-59-1, Fuxing Road in Beijing, the Communist Party of China History Memorial Hall in Luoyang, Chinese Embassy in Estonia (Tallinn), etc. The renovation and utilization of historic heritage buildings prolong their life cycle to achieve building carbon reduction goals, and create contemporary new buildings that coexist with old ones.

二、旧城疏解改造"分形加密、重建规制"设计方法

针对存量城市发展与更新中旧城疏解改造的空间密度和秩序问题，问题，建立了"分形加密、重建规制"的可持续设计方法。强调对旧城环境及肌理结构的保护延续和规制再生，改善空间质量、传承历史风貌、进化城市文明。

北京大院胡同28号改造工程[1]，位于北京市旧城西单地区丰盛胡同区域。设计采用了"分形加密、重建规制"的方式（图5-5），线形混凝土结构/空间单元构成了内含于整体院落群组建筑的空间架构，内含服务空间、形成主

[1] 获得WA中国建筑奖居住贡献奖优胜奖、中国建筑设计奖（历史文化保护传承创新类）二等奖。
Won Residential Contribution Award of WA China Architecture Award, the second prize of China Architecture Design Award (Historic and Cultural Protection, Inheritance, and Innovation Category).

II. The design method of using "Type encryption and morphology – reconstruction" to relieve and transform old urban contexts

In response to the spatial density and order issues in the downzonihg and transformation of old urban areas in the development and renewal of existing cities, the sustainable design method of "fractional encryption and morphology reconstruction regulation" was established, which emphasizes the protection, continuation, and morphology regeneration of the old urban environment, fabrics and structure, a facilitates improvement of the quality of space, inheritance of historic features, and evolvement of urban civilization.

Renovation of No. 28 Dayuan Hu-tong[1] is located in the Xidan-Fengsheng Hutong area of the old city of Beijing. The design adopts the method of "type encryption, and morphology reconstruction" (Fig. 5-5). The linear concrete

体支撑结构，又成为联系居宅和庭园的入口廊道，将大杂院转变为延续北京城市结构秩序的微缩社区，使人们既可重归各自日常宅园的诗意生活（图5-6），又能登高共享体验老城的都市胜景，以个案回应了北京旧城更新中人口密度、生活质量和风貌传承这"三道难题"，成为北京旧城进化更新的实验性范例之一。

　　该设计方法还被应用于北京护国寺西巷37号院改造——叠合院、北京隆福寺地铁（东四站）商业文化综合体等多项旧城疏解改造建筑工程。

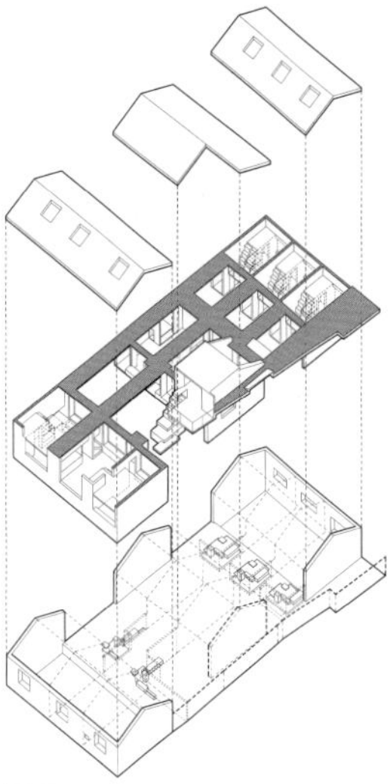

图 5-5

图5-5 "分形加密、重建规制"的设计策略
Fig. 5-5 The design method of "type encryption, morphology reconstruction"
图5-6 日常宅园的诗意生活
Fig. 5-6 The poetic life of daily homes

structure and space units constitute the space structure in the whole courtyard group, which contains service spaces, forming the main support structures and becoming the entrance corridors to be connected with the residences and the courtyards. The design transforms a mega-family into a miniature community that continues the order of Beijing's urban structure so that people can return to the poetic life of their daily homes (Fig. 5-6) and climb up and enjoy the urban scenery of the old city. The building responds to the "three problems" of population density, residential life quality and inheritance of traditional appearance in Beijing's old city renewal with singular case. It has become a practical example of Beijing's old city's evolution and renewal.

The design method has also been applied to the renovation of No. 37 Huguosi west lane - Congruent Courtyard House, the commercial block of Longfusi Station of Line 6 of Beijing Subway and other old city transformation projects.

图 5-6

图 5-7

三、居住建筑"理想空间、当代重构"设计方法

　　针对城镇发展中居住建筑空间品质提升和传统构型再生的问题，建立了"理想空间、当代重构"的可持续设计方法，重构具有地域性内涵的当代居住空间，探索平庸高密度城镇环境中的理想居住模式及居住空间的个性化和体验性特征。

　　唐山第三空间综合体工程[1]，位于河北省唐山市中心地区建设北路，紧邻一片工人住宅（图 5-7），建筑朝向、布局和塔楼及裙房的体量、形状几乎完全由日照计算得出，以满足严格的日照法规要求。在快速、简单和人工化、平

图 5-7　唐山第三空间综合体工程航拍
Fig. 5-7 Aerial image of the "Third Space" complex in Tangshan

[1] 获得中国建筑学会建筑创作大奖（2009-2019）、全国优秀工程设计行业二等奖、WA 居住贡献奖优胜奖。
Won the Architectural Creation Award (2009-2019) of the Architectural Society of China, the Second Prize of the National Excellent Engineering Design Industry, Contribution Award of WA Housing Award.

III. The design method of creating "ideal space and contemporary reconstruction" for residential architecture

In response to the improvement of residential building space quality and the regeneration of traditional configurations in urban development, a sustainable design method of "ideal space and contemporary reconstruction" was established. Which reconstructs contemporary living spaces with regional connotations, and explores the ideal living mode in the mediocre high-density urban environment and the personalized and experiential characteristics of living spaces.

The "Third Space" [1] project is located on Jianshe North Road in the central area of Tangshan, Hebei, next to a cluster of worker's residential buildings (Fig. 5-7). The orientation, layout, mass, and shape of the towers and podiums are almost entirely calculated from sunlight to meet the strict regulation. In the fast-paced, simple and artificial, banal urban environment of the

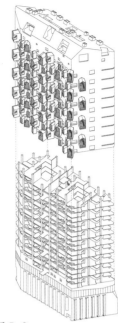

图 5-8

庸化的震后重建城市环境中，设计探索了居住单元空间的个性化特征，将通常标准层的平直楼板改变为层层堆叠的错层结构，形成居住单元中连续抬升的地面标高，犹如几何化的人工台地（图5-8），赋予了居住生活的多样体验，建成了两座包含76套"别业"宅园向高空延伸的立体城市聚落，实现了城市高密度住宅中的"理想居住"（图5-9）。

　　成都安仁里居住小镇工程，位于四川省成都市大邑县安仁古镇镇郊区域。设计以廖维公馆为核心和出发点，以林盘聚落作为类型参照，以其格局所在的轴线向四周延展，形成建筑、街道、田埂、水系与植被。较高的单元式住宅和商业建筑坐落于地块四周，环抱中心低矮的合院与叠拼

图 5-9

post-earthquake reconstruction, the design explores the personalization of the living units. It has changed the standard floor into a layered, staggered structure forms with continuously rising levels in the living unit, just like artificial geometric platforms (Fig. 5-8). The design not only endowed variety to the living experience but also built two three-dimensional urban settlements with 76 sets of "Bieye" (villa) residential courtyards extending to the air, realizing an "ideal living" mode in the urban high-density residential buildings (Fig. 5-9).

Anrenli Community is located in the suburban area of Anren Ancient Town, Dayi County, Chengdu. By taking the Liao Wei Mansion as the core and starting point, the design uses the Linpan settlement as the prototype. It expands to the surrounding from the axis of the masterplan, forming buildings, streets, ridges, water systems, and vegetation. The higher residential units and commercial buildings are located around the site, enclosing lower courtyards

图 5-8 居住单元空间的个性化特征
Fig. 5-8 The personalization of the living unit
图 5-9 城市高密度住宅中的"理想居住"
Fig. 5-9 An "ideal living" mode in the urban high-density residential buildings

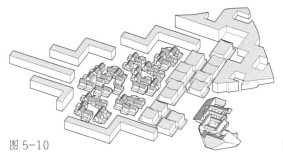

图 5-10

图 5-11

图 5-10 成都安仁里居住小镇工程轴测图
Fig. 5-10 Axonometric drawing of Anrenli Community
图 5-11 "林盘"式邻里生活空间构型
Fig. 5-11 The unique "Linpan" neighborhood spatial configuration

别墅，穿行其间的街道、田埂联系了"彼时"的林盘公馆和"此时"的"林盘小镇"与城镇公共空间（图 5-10），共同描摹了一幅汇聚了安仁聚落文化特质的、寄情田园农耕的生活场景，完成了对地域环境特有的林盘式邻里生活空间构型的当代延续和重构（图 5-11）。

该设计方法还被应用于南京安品园舍、西柏坡华润希望小镇等多项新建城镇居住建筑工程。

四、乡村建造"空间记忆、在地建造"设计方法

针对乡村建造中在地性与现代性营造之间的矛盾问题，

and stacked villas inside. The streets and fields crossing among the courtyards attach the "original" Linpan mansion to the "contemporary" Linpan neighborhood and the public space of the town. All of these describe a life scenery that contains the cultural characteristics of the Anren settlement and the spiritual yearning for pastoral farming (Fig. 5-10), completing the contemporary continuation and reconstruction of the unique "Linpan" neighborhood spatial configuration of the regional environment (Fig. 5-11).

This design method has also been applied to other new residential construction projects in cities and towns, such as Anpin Garden Houses in Nanjing and Xibaipo China Resources Hope Town.

IV. The design method of inheriting "spatial memory and local identity" for rural development

In response to the contradiction between territorial and modernity in

建立了"空间记忆、在地建造"的可持续设计方法。依托地理环境特征和空间建造基因，探索在土地中自然生长的当代乡村建筑的现代性特征。

　　楼纳露营基地服务中心工程，位于贵州省兴义市东部山区的楼纳村大冲组"建筑师公社"——群山环绕下的一块闭合盆地之内。设计结合了当地特有的喀斯特地质地貌特征以及场地中原住居民建筑拆除后遗留的房基和残墙、水井，把新建筑视为老宅基因的延续，新的功能由院落的方式组织，并向山林开敞，石阶将地面的多样活动引向连接成一体的建筑屋面，利用并改良了当地的石砌＋混凝土

rural development, the sustainable design method of "spatial memory and local identity" was established, which based on the characteristics of the geographical environment and spatial construction genes, explores the modernity characteristics of contemporary rural architecture that naturally grows in the land.

Camping Service Center in Louna is located in the "Architect Commune" of the Dachong group of Luna Village in the eastern mountainous area of Xingyi City, Guizhou Province – an enclosed basin in the mountains. The design combines the unique local karst geomorphological features and the ruined house foundations, residual walls, and water wells from the demolition of the aboriginal buildings on the site. It regards the new building as a continuation of the old genes. The new functions are organized with courtyards and open to the mountain forest. The stone steps lead various activities on the ground to the integrated building roof, using and optimizing the local

图 5-12

混合建造体系（图5-12），整个建筑犹如巨石匍匐于馒头山脚，与楼纳的独特地景融为一体，在地理环境、空间记忆和在地建造三个层面上作出回应，探索了一种包含隐喻的、在土地中自然生长的现代性（图5-13）。

　　该设计方法还被应用于楼纳大冲组民宿改造等多项乡村建造工程。

图 5-12 楼纳露营基地服务中心工程鸟瞰
Fig. 5-12 Overlook of the Camping Service Center in Louna
图 5-13 延续老宅基因的设计策略
Fig. 5-13 Design methods of maintains the old genes

stone + concrete hybrid construction system (Fig. 5-12). The whole building, integrated with the unique landscape of Louna, is like a boulder crawling at the foot of "Mantou (steamed bun) Mountain", responding to the surroundings at three levels: geographical environmental, spatial memorial, and local constructional, to explore metaphorical modernity that grows naturally in the land (Fig. 5-13).

This design method has also been applied to other rural development projects, such as renovating homestays in the Dachong Group in Luna.

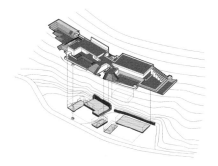

图 5-13

"生态、有序、平衡"的人居胜景
"Sustainable, Orderly, Balanced" human settlements

结语

关于工程建筑学

Epilogue

About

Engineering-Integrated Architecture

工程建筑学的提出，无意建立一个新的学科来否定和替代传统建筑学，而是对经典建筑学、建筑设计及建筑师职业本质的回归，是对现代以来传统建筑学的某种拓展和升级，希冀在当代人类赖以生存的生态、技术和空间环境发生巨变的状况下，尝试使经典建筑学更加具有系统性和科学性，以应对社会、经济、文化发展所带来的史无前例的学科困境和挑战，真正实现建筑学中艺术和科学的统一，突破当代生态、技术、环境难题，努力再造"生态、有序、平衡"的人居胜景，服务于国家高质量发展之战略和人类生境之未来。

The proposal of Engineering-Integrated Architecture does not intend to establish a new discipline to challenge or replace traditional architecture, but more like backtracking to the classical architecture, architectural design, and the essence of the architect's profession, becoming a specific expansion or upgrade of traditional architecture in modern times. With dramatic changes in the ecological, technological, and spatial environment on which contemporary human beings depend, it tries to make classical architecture more systematic and scientific to cope with the unprecedented disciplinary dilemmas and challenges brought about by social, economic, and cultural development, to realize the unity of art and science in architecture. Meanwhile, by tackling the contemporary ecological, technological, and environmental problems, it is striving to recreate the "Sustainable" "Orderly" "Balanced" human settlements, and serving the country's high-quality development strategy and the future of human habitats.

"生态、有序、平衡" 的人居胜景
"Sustainable, Orderly, Balanced"
human settlements

附录

Appendix

建筑学中的工程、技术与意匠[1]
李兴钢、郭屹民、张准、郭建祥、袁烽、
林波荣、祝晓峰、章明、罗鹏

　　李兴钢：工程建筑学的具体含义是指建筑技术与艺术的融合或者相互的启发。这个思想的提出并不是我从零创造发明的，而是受我们这个学科的前辈们思想和设计实践的启示。建筑学是一门古老的学科，其创立之初即是建筑与技术融合的状态，包括建筑师的职业工作也是将与技术有关的创作和创新密切地统合为一身的。随着人类科技的

[1] "工程建筑学"专辑9人对话（2021年8月25日）发表于《当代建筑》2021第10期，此处有较大删减和修改。The original text was published in the "Engineering-Integrated Architecture" album of the 10th issue of *Contemporary Architecture* 2021, which has been greatly modified here.

"Engineering-Integrated Architecture" album of 9-person discussion[1]

Engineering, Technology and Ingenuity in Architecture
Li Xinggang、 Guo Yimin、 Zhang Zhun、 Guo Jianxiang、 Yuan Feng、
Lin Borong、 Zhu Xiaofeng、 Zhang Ming、 Luo Peng

Li Xinggang: Engineering-Integrated Architecture refers to the integration or mutual inspiration of technology and art in architectural design. The idea was not invented from scratch by me, but was inspired by the thinking and design practice of our predecessors in this discipline. Architecture is an ancient discipline that was founded in a state of integration of architecture and technology, including the professional work of architects, which also closely integrated technology-related creation and innovation. With the

发展，建筑规模越来越大、功能越来越复杂，产生了工程技术学科与建筑学科的分离与相互独立，虽然这是时代发展先进性的一种体现，但也不可避免地带来了两类专业在建筑设计中必须思考的内容一定程度上的分离。建筑设计所要服务的社会环境、空间环境越来越复杂，越来越具有挑战性，在某种程度上建筑与工程的分离也出现了与之不相匹配和适应的状态，给我们的行业及学科带来了困扰。当代建筑创作中出现不尽如人意的现象，比如对建筑艺术属性的过分凸显或者个人表达主观随意性的凸显；比如对于科学技术属性的过度强调或者建筑与人的生活和情感相

development of human science and technology, the scale of buildings has become more extensive, and the functions more complex, which has resulted in the separation and independence of engineering disciplines from architectural disciplines. Although this is a manifestation of the advanced nature of the era, it also inevitably brings certain extents of separation that the two types of professionals must think about in architectural design. The social environment and spatial environment that architectural design is to serve is becoming more and more complex and challenging. To some extent, the separation of architecture from engineering has also appeared in a state of incompatibility and in-adaptation, which has brought trouble to our industry and disciplines. Unsatisfactory phenomena appear in contemporary architectural creation, such as an excessive prominence of architectural and artistic attributes, or a prominence of the subjective arbitrariness of personal expression, an overemphasis on the attributes of science and technology, or

疏离的状态，都可不同程度反映工程和建筑学分离所产生的问题。这也是建筑的工程、技术和艺术创作相融合，或者"以工程引导意匠""将技术转化为诗意"——工程建筑学思想的提出背景。

郭屹民[1]：在建筑技术不断发展的今天，来讨论"工程""技术"与"意匠"的问题，在我看来是非常有必要和及时的，并且能通过这样的讨论，让我们来重新认识在建筑学层面上的那些"工程"与"技术"的含义。按照渡边邦夫（Kunio Watanabe）的观点，"技术是科学这

[1] 郭屹民：东南大学建筑学院副教授。
Guo Yimin: associate professor, School of Architecture, Southeast University.

an alienation of architecture from human life and emotions. All these reflect the problems arising from the separation of engineering from architecture to varying degrees.This is also the background for the integration of architectural engineering, technology and artistic creation - or the idea of "engineering-led ingenuity" and "technology transformed to poetics" - Engineering-Integrated Architecture.

Guo Yimin[1]: In today's continuous development of construction technology, it seems to me that it is very necessary and opportune to discuss the issues of "engineering" "technology" and "ingenuity", and through such discussions, let us revisit the meaning of "engineering" and "technology" at the architectural level. Kunio Watanabe states, "Technology is the concrete means of science as an abstract concept, while 'engineering' is making those incomprehensible abstract sciences into usable, operational forms of matter, through the translation of means of technology. From an etymological point

一抽象概念的具体手段，而'工程'则是通过'技术'手段的翻译，将那些人们无法理解的抽象科学变为可以使用的、操作的物质形式"。从词源的角度，"工程"的英文是engineering，它来自于engine，而engine则可以追溯到拉丁语ingenium，本意是"内在的才能"，而ingenium跟英文中的ingenious、ingenuity这些表示心灵手巧、独创性的词都是同源。"技术"，也就是technology，它的词源来自于拉丁语的technologia，或tekhne，最初也是指处理艺术或工艺对象时的技巧，英文应该是technique。这种个人化的技巧被定型化、经验化之后，就形成了更为教条的

of view, the English word of "engineering", comes from "engine", and "engine" can be traced back to the Latin word "ingenium", which initially means "inner talent". "Ingenium" and the original words of "ingenious" and "ingenuity" are homologous. "Technology" is derived from the Latin word "technologia" or "tekhne", which originally also referred to the skills when dealing with artistic or craft objects, so-called "technique". After this personal technique became stereotyped, a more dogmatic "technology" was formed. The two words "engineering" and "technology" in the context of our cognition today have obviously lost the meaning of "cleverness" such as "ingenuity" and "skill", but instead more directed to science-related external reference. They are very different from the intuitive original meaning, which originated from the human mind and intuition. In the prologue of *Studies in Tectonic Culture: The poetics of Construction in Nineteenth and Twentieth Century Architecture*, Harry Francis Mallgrave makes it clear that tectonic "presupposes human

technology。"工程"与"技术"这两个词在今天我们认知的语境之中，显然已经完全没有了"心灵手巧""技巧"之类的"巧妙"的含义，更多地被指向与科学相关的外部关照。它们与"工程""技术"最初源自于人类心灵的、直觉的本意显然是大相径庭的。在《建构文化研究》的序当中，马尔格雷夫（Harry Francis Mallgrave）就明确指出了建构"是以人类对世界的心理和情感认知为前提的"。因此，不是什么建筑的技术都可以冠以"建构"的名目，技术需要以人类学的关怀来呈现，才能真正属于建筑。

二十世纪五六十年代的"结构表现主义"曾经被认为

psychological and emotional cognition of the world." Therefore, not all architectural technologies can be titled "tectonic", and technologies must be presented with anthropological concerns to belong to architecture truly.

The "Structural Expressionism" of the 1950s and 1960s was once considered a symbol of the perfect combination of architecture and structure. To this day, this symbol still seems to be used as a symbol, or a limitation that has become a "ceiling" for the integration of architecture and structure. However, when you see that many works of "structural expressionism" are dominated by structural designers, you will realize that I want to emphasize the intention of building technology from an architect's perspective.

The different ways of structural expression are generally described, such as structural performance, decoration, images, meaning, and other different forms of expressions. From a complete objectification, to objectified visual

是建筑与结构完美结合的一种象征。至今这种象征似乎仍然在被作为一种符号，或者说是一种局限，成了建筑与结构相互融合的"天花板"。不过，如果大家能够发现"结构表现主义"很多作品都是结构设计师们主导的作品时，就会意识到我想强调的是站在建筑师视角的建筑技术的意图了。

对于结构表现的不同形式进行大体上的描述，比如结构的性能、装饰、图像、意义等不同的表现形式。从完全客观化，到对象化的视觉形式，到主观感受的知觉体验，可以说在与身体关系的各个层面上，结构都可以纵横驰骋其中。对于结构的认知完全可以被看作是对建筑的理解。

forms, to the perceptual experience of subjective feelings, it can be said that at all levels of the relationship with the body, the structure can run in all direction. The perception of structure can be seen as an understanding of architecture. Compared with the performance of the structure and the expression of force flow, the "ingenious" structural expression related to human intelligence is full of wit and humor. The observer can empathize with the designer, and obtain the common emotions of human beings with joy, anger, sorrow, and happiness. This structure is capable of having a human presence beyond structural technology. It may be closest to our body in the structural representations we know, or the most architectural one.

与结构的性能、力流的表达相比，充满着机智、幽默的，与人类智慧相关的巧妙的结构表现中，观察者可以与设计者感同身受，获得人类的共通情感，喜怒哀乐。这种结构能够具有超越结构技术之外的存在。它可能是我们所认识的结构表现中距离我们身体最近的一种，或者说是最建筑的一种吧。

李兴钢："技术需要以人类学的关怀来呈现，才能真正属于建筑"，从结构工程师或者土木建筑学科的角度来讨论工程建筑学，与建筑师讨论的话题和尺度会有所不同。

Li Xinggang: "Technology must be presented with anthropological care to belong to architecture truly". If Engineering-Integrated Architecture is discussed from the perspective of structural engineers or civil engineering disciplines, the topics and scales will be different from which architects discuss.

张准[1]：如果单纯从结构角度考虑，安全性和经济性是最核心的内容。以科学性态度去理解、研究、教授结构，对结构学科发展是必要的，但在实践中结构也可以是个很泛化的概念，除了与力学相关的、和安全经济性关联的结构之外还有很多抽象的结构或者可以用结构化思维方式对待的事物。这些泛化的结构概念蕴藏了很多可能性，可以与建筑师的设计结合到一起，又不失结构逻辑性，也容易被结构师理解。把建筑师的创意从设计意象或人文语境，转移到结构设定的语境上，但这中间需要大量的沟通、磨合、试错。

[1] 张准：和作结构建筑研究所联合创始人。
Zhang Zhun: co-founder of Hezuo Structure Architecture Research Institute.

Zhang Zhun[1]: If considered purely from a structural point of view, safety and economy are the core contents. Understanding, researching, and teaching structure with a scientific attitude are necessary for advancing the structure discipline. However, in practice, structure can also be a generalized concept, and there are many abstract factors or things that can be treated with a structural methodology, in addition to those related to dynamics, safety, and economics. These generalized structural concepts contain many possibilities, which can be combined with the architect's design without losing the structural logic, and also easily understood by the structural engineer. The architect's creativity is transferred from the design image or humanistic context to the context of the structural setting, but this requires a lot of communication, coordination and trial-and-errors.

In the process of collision and dialogue, some docking tools are needed: graphical statics, and parametric tools etc., or other methods that can produce

在碰撞与对话的过程中，需要一些对接的工具：图解静力学、参数化工具等等，能将结构的力转化成视觉的表达，或者其他能产生直观形态表达的方式都可以，即使最传统的手工模型和草图也能够弥补中间环节的缺失。此外，与建造逻辑相关的施工工法、工艺，甚至具体到技术加工细节衍生出的一套逻辑，也可以用来弥补中间的空白。如果建筑师与结构工程师之间长期合作，比较容易深层感知对方想要表达什么，能在更为宽阔的语境下的彼此沟通对话，会超出技术工具和工法的层面。

结构介入的时间点很关键，但无论结构介入的早晚，

intuitive morphological expressions, or that can convert the force of the structure into visual illustrations. Even the most traditional manual models and sketches can compensate for the lack of intermediate links. In addition, construction methods and craftsmanship related to the construction, and even a set of logic derived from the details of technical processing can also be used to make up for the gaps. If there is long-term cooperation between architects and structural engineers, it is easier to perceive clearly what the other part wants to express. When they communicate and dialogue with each other in a broader context, it will go beyond the level of technical tools and construction methods.

The timing of the structural intervention is critical, but whether sooner or later, the architect always plays a decisive role in this interaction. Architects and structural engineers should have the knowledge reserve of the other part as much as possible. If both parties can have an equal

建筑师在这个互动中都起到决定性作用。建筑师和结构师也要尽可能有对方的知识储备，如果双方能有对等的觉悟，去建议与发展这个过程，就很容易找到契合点，引发出不一样的新东西，而且那条线索会深入建筑的精髓所在。

李兴钢：建筑师只有理解结构、受力、建造逻辑以及规范方面的内容，才能够更好地与结构工程师进行沟通合作。换个角度，建筑师如果能够激发出结构工程师对建筑师设计意图的理解，或者创造性的理解，才能够让结构工程师更加主动地帮助建筑师实现他的设计意图。项目开始

awareness to recommend and develop this process, it is easy to find a point of convergence, and trigger a different new thing, and that clue will go deep into the essence of architecture.

Li Xinggang: Architects who understand the logics of structure, force, construction, and specifications can better communicate and collaborate with structural engineers. From another perspective, if the architect can stimulate the structural engineer's understanding of the architect's design intention or creative ideas, the structural engineer can be more active in assisting the architect in realizing his intention. At the beginning of a project, the architect's design intention may not be clear; in other words, it may not be the best solution at this stage. Thus the "creative" understanding of the structural engineer is critical. In order to achieve an optimal integration of architecture and structure, the timing when architects and structural engineers intervene is essential to enter a state of common thinking

之初，建筑师的设计意图可能并非清晰，换句话说也许并不是最佳的解决方案，此时结构工程师的"创造性"理解就非常的重要。建筑和结构要想做到最佳的融合，建筑师与结构工程师介入的时间点非常重要——从开始就进入共同的思考和创作的状态；同时双方要有同理心，对对方的工作要有充分的理解。关于建筑师和结构工程师以及设备工程师的合作，在机场等大型项目中会经常遇到非常复杂的技术配合情况，那么究竟是什么样的合作基础能产生好的合作结果和作品呢？

and creation from the beginning. Simultaneously, both sides must have empathy and a complete understanding of each other's work. Regarding the cooperation between architects and structural engineers and equipment engineers, in large projects such as airports, there are often very complex technical cooperation situations. So what kind of cooperation basis can produce good cooperation results and works?

郭建祥[1]：追溯建筑学发展的历史，建筑与结构最初是合二为一的，建筑师既要负责建筑创意的创作也要负责建造实施的落地，实现两者完美融合的工程作品。后续随着技术、时代的发展，工程项目中的专业划分越来越细，从建筑学、结构再到机电设备乃至绿色、消防等方面，各种各样的技术在同步推进着工程技术的发展。事实上，各个专业应该统一在同一逻辑下，围绕同一根技术主线交织在一起共同完成，而不是两条平行线或者说两条仅有一个交点的交线。

[1] 郭建祥：全国工程勘察设计大师，华建集团华东建筑设计研究总院副院长、首席总建筑师。
Guo Jianxiang: honorary winner of National Engineering Survey and Design Master, vice president and chief architect of East China Architecture Design and Research Institute of Huajian Group.

Guo Jianxiang[1]: Tracing the history of the development of architecture, we can see that architecture and structure were initially integrated as one, and the architect was responsible for both the creation of the architectural idea and the realization of construction and implementation, so as to achieve engineering works of a perfect integration of the two. With the development of technology and times, the division of expertise in engineering projects has become more detailed. Various technologies are advancing engineering technology, from architecture and structure to mechanical and electrical equipment, and even green architecture and fire fighting. In fact, all disciplines should be unified under the same logic, intertwined around the same technical main line, rather than two parallel lines or two lines with only one intersection.

At present, there are two trends in architectural creation: one is to exaggerate the creativity, shape, and the skin out of the technical level,

目前建筑创作中存在两种趋势：一种是将创意、造型、表皮脱离技术层面而夸张化，将其变得特别复杂、繁复，建筑设计在整个工程中与技术是脱节的，虽然目前技术水平的发展（例如参数化设计、大跨度技术）完全可以支撑任何建筑的相对夸张的形态，但过于夸张的造型带来了过度表达的问题，并影响造价和功能合理性；另一种情况，就技术这条主线而言，各种技术存在于同一维度中共同作用于建筑创作，其中结构技术对建筑师的创意影响最大。目前我感到抗震技术、消防技术对于超大型项目的建筑设计也会产生很大的约束。越来越严格的消防规范对空

resulting in unnecessary complications. Architectural design is disconnected from technology throughout the construction process. Although development on the technical level can fully support any exaggeration of forms building (such as parametric design and large-span structure), excessive forms bring about problems of over-expression and affect the economic and functional rationality. In the other trend, as far as the main line of technology is concerned, various technologies exist in the same dimension and act together in architectural creation, of which structural technology has the greatest influence on the creation of architects. At present, I feel that seismic and fire fighting technologies will also have great limitations on the architectural design of super-large projects. Increasingly stringent fire codes also have a significant impact on spatial creation and so on. Other related techniques limit architectural creation to varying extents. In the face of such a situation, in the creation process, we must be able to understand the limitations

间创意等方面也会带来很大的影响。其他相关的技术都会在不同程度上限制建筑创作。面对此类情况，在创作过程中，一定要能够理解相关技术对建筑的创意限定，要把握住这根主线来逐步完成从创意到实施的过程，这应该是建筑师需具备的基本能力。

因此，不同工种的工程师应理解建筑创作过程中的相应创意逻辑，在不同的设计阶段，利用最合适的技术、最合适的对应方法来实现创意的落地。在方案创意中，结构工程师要参与进来，对整个创意的走向有基本判断并提供技术层面的可行性支撑，从而构成方案实施的基础；接下

of relevant technologies on building, and we must grasp this main line to gradually complete the process of this creativity to implementation. This should be a basic ability of architects.

Therefore, engineers of different disciplines should understand the corresponding creative logic in architectural creation, and use the most appropriate technology and corresponding methods to realize creative achievement at different design stages. During the Schematic Design stage, the structural engineer should participate to have an essential judgment on the direction of the whole idea and provide feasible technical support, thus forming a basis for the implementation of the plan; In the following stage of Design Development, more emphasis is required on the systematicness of technology to realize the landing of the creativity. Under the premise of the primary architectural functions, all technical disciplines complete the design together around the same design principle to achieve the best adaptability

来的扩初阶段要更多地强调技术的系统性来实现创意的落地性，各个技术专业在以建筑基本功能为前提的基础上，围绕同一设计原则来共同完成设计，将其做到适配性、匹配度最好；在施工图阶段，建筑师更多的注意力则是在细节把控上，通过建筑语言完成文化性的表达，通过技术的细节表达将建筑创意品质完整落地，实现较为完美的建筑品质，这正是我们以技术条线为主来做的工作。此外，我们还要避免过度设计的情况发生。最后，我想强调的是"平衡"。总体而言整个建筑工程中一定要在艺术与技术的结合中找到一条主线，通过交织、交融、平衡的方式将建筑创

and matching degree; In the Construction Documentation stage, architects should focus more on detail control, complete cultural expression through architectural languages, and realize the creative quality through the complex expressions of technology to achieve perfect architectural quality, which is precisely what we do with technical clues. In addition, we should also avoid over-design. Finally, I would like to emphasize "balance". In general, the entire architectural project must find a main line in combining art and technology, and eventually evolve architectural creativity into high-quality constructed works through interweaving, blending, and balancing.

意最终演变为高品质的落地作品。

李兴钢：大项目的难度和复杂性在我们今天话题的讨论中会有非常不一样的状态。在越来越复杂的技术限定条件下，我们怎样能够通过寻找到一条主线，然后跟各种副线交织、交融，最后用系统性的技术方式，创作出具有平衡性、合理性及适配性的创作结果。现在技术的发展中新的技术和复杂技术的重要代表是参数化建造（数字化设计）。

袁烽[1]：今天的话题非常及时，也能够对中国当下的建

[1] 袁烽：同济大学建筑与城市规划学院教授，博士生导师。
Yuan Feng: professor and doctoral supervisor of School of Architecture and Urban planning, Tongji University.

Li Xinggang: The difficulty and complexity of large projects can be very different in our discussion of today's topic. Under the increasingly complex technical constraints, how can we find the main line, interweave and blend with various sub-lines, and finally use systematic technical methods to realize balanced, rational and adaptable creative outcomes. An important representative of new and complex technologies is parametric construction (digital design).

Yuan Feng[1]: Today's topic is very timely and can also have a leading impact on the current architectural practice in China. The Engineering-Integrated

筑实践现状作出非常有引领性的影响，兴钢总提出的工程建筑学，对日后我们理解建筑学，理解建筑和工程的关系，有着非常重要的历史价值和时代意义。Engineering（也被称为 design engineering），翻译过来就是工程学或者设计工程学，首先它是一个非常跨学科的内容，追溯到文艺复兴时期，当时阿尔伯蒂在《论建筑》中提出前卫的观点：设计师不建造，建造师不设计。之所以前卫是因为文艺复兴之前的设计者、建造者，以及绘画者和雕塑者都是同一职业。但是从文艺复兴开始，在阿尔伯蒂《论建筑》之后，设计师和建造者就是分开了，至今也没有合并。在今天提

Architecture proposed by Li Xinggang has a very important historical value and contemporary significance on our understanding of architecture and of the future relationship between architecture and engineering. Engineering is also known as design engineering. First of all, it is a very interdisciplinary content, dating back to the Renaissance, when Alberti put forward the avant-garde view in *On Architecture* (*De re aedificatoria*): "Designers don't build, builders don't design." It is avant-garde because the designers, builders, painters and sculptors before the Renaissance were all of the same profession. But from the Renaissance onwards, after Alberti's *On Architecture*, the designer and the builder were separated and have never merged again. Today, I mentioned the new integration – the integration of architecture and engineering, breaking the original architectural ontology that iterates with style and re-examining a turning point beyond the time. This is why I support Li Xinggang to propose the era of integrating architecture and other related disciplines.

到新的融合——建筑与工程的融合，打破原有以风格来迭代的建筑本体学的提法，重新审视了一个超越于时代的转折点，这是我非常支持兴钢总提出的建筑和旁支学科交融的时代的一点。

回到国际化的视角来看待工程建筑学，engineering 的提出不仅仅是在中国，design engineering 的话题在西方语境下"融合"也是近十年非常注重的话题，而且呈现出新的趋势——关注数字化和engineering 的关系。建筑学科因为数字时代的到来打通了新的领域，engineering 的理解不仅仅限于结构，还有消防、环境性能，这些专业都

Returning to the perspective of internationalization to examine Engineering-Integrated Architecture, engineering is not only proposed in China but also a topic of "fusion" in the Western context in the past decade, along with a new trend focusing on the relationship between digitalization and engineering. Architecture has opened up new fields with the arrival of the digital age. The understanding of engineering is not limited to the structure but also fire fighting and environmental performance, which are also related to the new integration with architecture. A distinctiveness of the new era is that all disciplines and tools take on the characteristics of instrumentalization. This instrumentalization is the new copyright—a shift from "starchitects" to a more complex creative autonomy that emphasizes collaboration between humans and digital tools. One of its features is the computational formal design, and the other is the computational rationalization design. Computational form design can also be called post-rationalization, through

在和建筑谈新的融合。新时代的特征是所有学科和工具都呈现工具化的特征。这个工具化就是新的著作权——从明星建筑师转向一种更加复合的强调人与数字工具协作的一种新创作自主权，它的特征一个是计算形式设计，一个是计算合理化设计。计算形式设计也可称为是后合理化，任何形式都能够通过后合理化的方式来帮建筑师实现创意。Computational design，在灵感和想象的同时加入合理工具来帮助建筑师进行形式的建造，也可称为生成式的或者是升级化设计。数字化著作权并不仅仅靠个人的智力，而会融入一些工具的合理化色彩，人和机器的这种比拼变成大

which any built form can be used to help architects realize their creativity. Computational design, which adds reasonable tools to help architects to construct forms with their inspirations and imaginations, can also be called generative design or upgraded design. Digital copyright does not only rely on individual intelligence but will be integrated into the rationalization of tools. The competition between people and machines becomes a stage of continuous research and development and continuous overcoming of stylization in which everyone confronts one another.

家相互对峙的一种不断研发和不断克服风格化的阶段。

　　未来的设计师能够在早期的教育中真正理解基本的学科知识，就需要所有的工具会被插件化，会被工具化。在高效能或者高性能的一个新时代提出工程建筑学，是与我们国家所强调的高质量发展、强调的更集约化发展是不谋而合的。对于建筑来说一个比较高的境界并不是炫技，而是如何能让技术变得看不见，隐藏在实际的内涵当中。未来如果我们能够为建筑行业带来看不见的技术，而且技术中的工具又没有被风格化的状态可能是我们向往的一种状态。

To truly understand the basic disciplinary knowledge in their early education, future designers will need all tools to become plugins and instrumentalized. Proposing Engineering-Integrated Architecture in a new era of high efficiency or high performance coincides with the more intensive development that our country emphasizes on high-quality development. A relatively high state for architecture is not a showmanship, but how to make technology hidden in the actual connotation. In the future, if we can bring invisible technology to the construction industry, and the tools in the technology are not stylized, it may be a state of our yearning.

李兴钢：当代最新的技术在engineering 方面的进展和目标，看不见的、隐藏的技术，让我回想起2015 年参加郭屹民老师组织一批日本建筑师和结构工程师在同济大学召开的国际结构建筑学研讨会，日本结构工程师大野博史提出来的一个概念就是"隐藏的技术"（Hide-tech）而非"高技术"（High-tech），强调最高级的技术并不是一种表现出来的状态。性能化设计，不光包含了结构性能化、消防性能化，还有非常重要的环境性能化。与建筑相关的环境调控技术在建筑设计中越来越重要，在中国追求高质量发展、高效能发展以及"双碳"目标提出的背景下，环境调控技

Li Xinggang: The progress and goals of the latest contemporary technologies in "engineering" are invisible and hidden technologies, reminding me of the International Symposium on Archi-Neering Design held at Tongji University organized by Professor Guo Yimin in 2015, with a group of Japanese architects and structural engineers participating. A concept proposed by Japanese structural engineer Hiroshi Ohno is "hidden technology" (Hide-tech) instead of "high technology" (High-tech), emphasizing that the highest level of technology is not a state of manifestation. Performance-oriented design are not only structural performance-based, and fire-fighting performance-based, but also, very importantly environmental performance-based. Building-related environmental control technology is becoming more and more crucial. In the context of China's pursuit of high-quality development, high-efficiency development, and the proposed "double carbon" target, the role of environmental control technology in architectural creation will become

术在建筑创作中的作用将会越来越凸显，越来越重要。将环境调控技术和建筑设计融合，是体现建筑中技术和创作关系的重要领域。

　　林波荣[1]：我觉得工程建筑学这个概念的提出非常及时，有两方面重要意义：第一，就是提醒我们不忘初心；第二，要紧跟时代的最新发展需求，体现时代特征。"工程引领意匠"这一建筑设计实践思想的归纳，是对行业发展和创新引领不可或缺的学术探索。

　　所谓不忘初心，早在1945年梁思成先生参与联合国

[1] 林波荣：清华大学建筑学院教授、博士生导师。
Lin Borong: professor and doctoral supervisor, School of Architecture, Tsinghua University.

more prominent. The integration of environmental control technology and architectural design is an important field that reflects the relationship between technology and creation in architecture.

Lin Borong[1]: I think the proposition of Engineering-Integrated Architecture is very timely, and it has two important significances: First, it reminds us not to forget our original intentions; second, we must keep up with the latest development needs of the time and reflect the characteristics of the time. The induction of the architectural design idea of "engineering-led ingenuity" is an indispensable academic exploration for the development and innovation of the industry.

The so-called not forgetting the original intention, dates back as early as 1945. when Mr. Liang Sicheng participated in the design proposal of the United Nations Building, he emphasized how the building layout is conducive

大厦设计提出的设计方案时，强调的是建筑布局怎样有利于通风，有利于利用自然环境条件，最后可以大幅降低空调能耗，他在那个时代就已经强调体形设计要有利于节能舒适，这种理念令人钦佩。兴钢总提出的工程建筑学理念，就是在提醒我们建筑学在作为一个高雅的艺术实践的同时，也应该强调工程实践如何坚实、充分地解决我国城市发展需求问题。

所谓时代特征，我们现在正身处于大力发展高质量、高效率、高品质的建筑，实现城镇绿色化、低碳化的时代。第一个转变，建筑行业的发展面临着从传统的以房屋

to ventilation and utilization of the natural environment, and eventually a significant reduction of the energy consumption of air conditioning. At that time, he emphasized that the body shape should be conducive to energy saving and comfort, which is admirable. Engineering-Integrated Architecture proposed by Li Xinggang is to remind us that architecture, as an elegant artistic practice, should also emphasize how engineering practice can solidly and adequately solve the problems of urban development in our country.

Talking about so-called characteristics of the time, we are now in an era of vigorously developing high-quality, high-efficiency, and high-standard buildings and realizing urban greening and low-carbonization. First of all, the development of the construction industry is facing a transition from traditional building-oriented to people-oriented industry. It is necessary to change from focusing on the evaluation of the building to the satisfaction

为本向以人为本转变。要从关注对房屋主体的评价，转变为对使用者需求的满意度评价，从 design by you、design for you 到 design with you 的转变，表明从一开始强调房屋简单而又被动地由我们人来设计，到后来强调人与建筑设计融为一体，揭示出我们对房屋的关注，从一个客观的关注转变为使用者角度的关注，这是工程建筑学里面非常关键的转变点，体现了时代特征；第二个转变，传统技术的堆砌无以为继，需要更加强调建筑师在设计中的主导作用，也就是强调气候的适应性。要用工程需求对传统的建筑创作进行客观约束，从形式的追求提升到科技工程原理

evaluation of users' needs, from "design by you" and "design for you" to "design with you", which shows a transition from the previous emphasis on a simple and passive design of the building, to the latter emphasis on the integration of people and architectural design. It reveals our concern about the building changing from an objective concern to a concern from the user's point of view. This is a critical transition point for the engineering-related architecture, reflecting the characteristics of the time. The second shift is that the staffing of traditional technologies is unsustainable, so it is necessary to emphasize the leading role of architects in design, as well as adaptability to climate. It requires objective constraints on traditional architectural creations from engineering requirements, from pursuing form up to pursuing scientific and technological engineering principles. Architects should also lead this process to do more with less, or it can only be half the work. The third shift emphasizes changing from a simple measure orientation

和科学的追求上，这也应该是建筑师主导的，才能事半功倍，否则只能是事倍功半；第三个转变，强调从简单措施的导向转变为性能效果的导向，而这种性能效果的导向可能就是工程建筑学的关键做法。应该鼓励用数据说话，根据工程上的问题作为建筑设计的导向。

这些转变对于在建筑设计主导下的多专业的技术创新协同，也应该提出新的要求。我们试图建立一种与传统先设计后性能优化不同的新工作流程，强调从人的需求、人的尺度来看待室内的环境营造，甚至强调用部分时间、部分空间的理念来开展创新设计。从初期建筑形体的方案优

to a performance effect orientation, which may be the critical practice of engineering-related architecture. It should encourage illustration by data and guide architectural design with engineering problems.

These transformations should also put forward new requirements for multi-disciplinary collaboration on technical innovation under the leadership of architects. We try to establish a new workflow that is different from the traditional process in which design goes first and performance follows, emphasizing the indoor environment creation from the perspective of human needs and scale. It even emphasizes the concept of using partial time and partial space to carry out an innovative design. From the optimization of the initial architectural form to the later optimization of the external envelope and the internal environment, and finally to people, realizing the integration of multiple spatial technology forms. Only with the common context and unified goal of Engineering-Integrated Architecture can we achieve the

化，到后期的室外围护结构与室内环境末端优化，最终到人，实现了多位一体的空间技术形态的融合。有工程建筑学这一共同的语境和统一的目标，才能够实现这种理念下的创新，不忘初心，紧跟时代脉搏，做出更多更好的建筑。

李兴钢：林波荣老师获得了 2020 年腾讯"科学探索奖"，是唯一一位建筑界获得此奖项的专家，他的工作非常重要，是将环境调控技术和建筑设计融合的重要领域，他不仅是这个方向的学术研究者，也是这个方向设计工作实践的专家。建筑中技术和创作的关系如此重要，很多当代

innovation under this concept, maintain the original intention, keep pace with the time, and make more and better buildings.

Li Xinggang: Mr. Lin Borong won the 2020 Tencent "Scientific Exploration Award", and he is the only expert in the architectural industry to receive this award. His work is significant, and in a vital field of the integration of environmental control technology and architectural design. He is not only an academic researcher but also an expert in practical work in this direction. The relationship between technology and creation in architecture is so essential that many contemporary architects have paid academic attention or even conducted specialized research, such as Zhu Xiaofeng, who has just completed his doctoral dissertation.

实践建筑师对这一领域进行了学术性关注甚至专门研究，比如祝晓峰建筑师刚刚完成的博士学位论文研究。

祝晓峰[1]：兴钢提出的工程建筑学对当代建筑学的发展来说有非常重大的意义。尤其是"工程"和"意匠"的关系，相信一定会对建筑师的设计思维以及建筑师和工程师的合作关系产生积极的影响。我在自己的设计中确实越来越重视技术与建筑、与环境的高度整合。我的论文《建筑作为人的延伸》的整体立论，是把建筑的演化看作是三件事情交替衍生的结果：第一个是身心的延伸，即身心需

[1] 祝晓峰：山水秀建筑事务所创始人。
Zhu Xiaofeng: founder of Scenic Architecture Office.

Zhu Xiaofeng[1]: The Engineering-Integrated Architecture proposed by Li Xinggang is of great significance to the development of contemporary architecture. In particular, the relationship between "engineering" and "ingenuity" is believed to positively impact architects' design thinking and the cooperative relationship between architects and engineers. At the same time, I have paid more attention to the profound fusion of technology with architecture and the environment. The overall argument of my dissertation *Architecture as an Extension of Man* is to view the evolution of architecture as an alternate derivation of three things: The first is the extension of the body and mind. That is, the needs of the body and mind are the original basis and subsequent engine of architectural design; The second is the extension of the ontology. That is, the architecture has its own intrinsic principle, and the ontology includes the spatial and tectonic forms. The third is the extension of the interaction. In contrast to the ontology of architecture, interaction is

求是建筑原初的基点和后继的引擎；第二个是本体的延伸，就是建筑有了自身的、内在的规律，本体包括了空间形制，还有建构形制。第三个是交互的延伸，相对建筑本体而言，交互是所有的外力，里面包含非建筑本体的、人的各种延伸系统，包含政治、经济、环境等，而其中占比特别大的部分就是技术。我认为这三件事情的共同作用，推动了建筑的不断演化，并且在不同的时代会有所侧重。

我特别想提出教育上的一种整合，特别是跨专业学习的整合。从本科教育开始，就要培养不同专业对其他相关专业的理解力，并通过协作寻找系统性的设计解答；另外

among all external forces, including various extension systems of the non-architectural ontology and people, such as politics, economy, environment, and a considerable part is technology. I think a combination of these three things has driven the continuous evolution of architecture, and has different focuses it in different eras.

In particular, I want to propose an integration in education, especially the integration of interdisciplinary learning. Starting from undergraduate, architects should cultivate an understanding of different related disciplines and find systematic design solutions through collaboration; In addition to professional education collaboration, it is also necessary to establish an overall understanding of architecture through the study of architectural manifestation. In fact, throughout history, architecture has been passed down as a traditional building system and craftsmanship. The complex requirements of buildings in terms of structural calculation, equipment, and

专业教育协作之外，还要通过对建筑形制的学习建立对建筑整体的认知。其实在历史上，建筑学就是作为一个传统的建筑体系和工艺被传承下来，建筑在结构计算、设备、消防等方面复杂的需求是工业革命之后才加入建筑设计工作中的，当代建筑给人类提供的服务不断膨胀，但其基本目的和历史相比并无很大差异，我们无非还是要让人待在一个安全的结构里，然后处于一个相对比较舒适的温度当中，从事工作、生活、社交等活动。那么在这个意义上，一种建筑形制或者历史上出现过的建筑原型，包括它包含的一些本质性的基本要素，这些都应该被结构、设备工程

fire fighting were only added to the architectural design after the industrial revolution. Although the services that contemporary architecture provides to human beings continue to expand, their primary purposes are not very different from those in history. We want nothing but people to stay in a safe structure and be in a relatively comfortable environment for work, life and social activities. In this sense, structural and equipment engineers should learn and understand a kind of architectural form or prototype that appeared in history, including some essential elements it contains. In the future education, we can have the opportunity to promote these to all relevant professionals

师所理解和学习。在未来的教育里能够有机会再把这些推广到所有从事建筑工程的相关专业人员，让各个专业的工程师和设计师都能够有一个讨论的基础，而且知道这个建筑作为一个原型，它是如何从历史慢慢演变到今天的，现代技术是如何融入或改变建筑类型的；还要知道我们现在作为人类整体，为什么需要这样的建筑，我们过去是怎么样，我们未来要怎么样？我们的建筑创作应该是一个长期的、整体的建构，亦是一种语境的建构。

engaged in construction engineering so that engineers and designers of various professions can have a common basis for discussion. Meanwhile, considering this building as a prototype, they can know how it has gradually evolved from history to today, and how modern technology has been integrated into or changed the type of building. They also need to know, as human beings, why do we need such buildings now, how we were in the past, and how we will be in the future. Our architectural creation should be a long-term, holistic, and also a contextual construction.

李兴钢：如何能够在与建筑相关的教育阶段，使将来的建筑师，以及所有参与到建筑设计工作的相关专业的工程师都能对建筑构成的基本要素有全面的理解？如何在对本质全面理解的基础上，能够实现在未来建筑创作中的技术运用？技术和设计融合的理念，在向更大的领域扩展——不光在建筑的领域，还在继续向城市基础设施的领域延伸，甚至拓展到风景园林的领域，以及工业遗产保护改造利用等方面。作为建筑教育工作者和学术带头人，对于这个话题的理解和行动，至关重要。

Li Xinggang: How can future architects, as well as engineers of all related disciplines, have a comprehensive understanding of the essence of architectural composition at the educational stage? How can the application of technology in future architectural creation be realized based on a comprehensive understanding of the essence? The integration of technology and design is expanding to a larger field, not only in architecture but also in urban infrastructure, and even in landscape architecture, and the protection, transformation, and utilization of industrial heritage as well. As educators and academic leaders, understanding and action on this topic is crucial.

章明[1]：工程建筑学的讨论代表着一种态度，或者代表着一种新的思考和新的视角。一个是反对过度的艺术化，另一个是反对过度的技术化，还要反对过度的装饰化，甚至可以说是反对跟技术逻辑没有关系的所谓过度精细化。由技术逻辑和艺术逻辑所综合的技艺逻辑，就以这种方式以及在这个时间点提出，具有超越时代的转折点的意义，我也原则认同。在大家做了大量相对盲目的甚至一些不同维度的实验之后，回归到一个相对理性地将技术与艺术相结合的理念当中，有助于它达到本体思想最契合的一种状态。回到建筑的整体过程当中，包括两个最核心的方

[1] 章明：同济大学建筑与城市规划学院教授、博士生导师，同济大学建筑设计研究院有限公司原作设计工作室主持建筑师。
Zhang Ming: professor and doctoral supervisor of School of Architecture and Urban planning, Tongji University, principal architect of Original Design Studio of Tongji University Architectural Design Institute.

Zhang Ming[1]: The discussion of Engineering-Integrated Architecture represents an attitude or new thinking and a new perspective. One such attitude is the opposition of excessive artistry; an other is the opposition of excessive technical appliction and excessive decoration or even the so-called extravagant refinement that has nothing to do with technical logic. The technical logic that synthesizes the logic of technology and that of art, proposed in this way and at this point in time, is significancant as the turning point of the times, and I am in favour of it. After a lot of attempts in various demensions, returning to a relatively rational concept of combining technology and art will help achieve a state that best fits the thought of ontology. Returning to the integrated building process, we can see that it includes two core aspects: structure and tectonic, or the shaping of form and space, and the expression of materials and tectonics. The concept of "framework" proposed by Kazunari Sakamoto in Japan is most differenent

面：结构和建构，或者说是形态和空间的塑造，以及材料和建构的表达。日本建筑师坂本一成提出的"架构"概念其实跟结构相比较最大的不同，就是指结合材料形成构筑物，还特别强调不仅是工程性的结构，而是让结构能够参与空间设计和场所塑造形成一个体系。其核心可以概括为：结构不是孤立存在的，而是以建筑本身的空间、形态、体验，甚至是与环境的关联共同表达在一起的，也就是把结构融入整个建筑体系当中。我们现在特别强调在设计过程中强化受力体系，通过对它的构件本身进行不同的职责分配，形成体系化的内容，服务于我们的形态和空间最终达

from structure in the sense of the combination of materials to form structures. It emphasizes not engineering structures but allows structures to be part of space design and place-shaping and form a system. Its core can be summarized as follows: The structure does not exist in isolation but is expressed together with space, form, experience, and even the association of environment with the building itself. The structure is thus integrated into the entire building system. We now emphasize strengthening the force system in the design process. A systematic content is formed by assigning different roles to its components, serving our form and space, and ultimately achieving a holistic effect.

到一个整体性的效果。

　　工程建筑学的提出可能会促进几个改变。一是教学的内容和模式的改变。我提出"融界"，是把界限打破，让它融合成为一个大的环境观，或者说一个大的营造观，能够把建筑学、城乡规划和风景园林都融为一个大的体系来思考问题，景观城市基础设施这个概念，不仅是有基础设施、建筑学，而且它还要成为城市公共空间中新的活动平台以及城市新的风景，在未来能够促进教学内容和教学模式的改变。二是要促进建筑师和工程师协作方式的改变，工程建筑学恰好为我们提供了一个依据。三是工程建筑学的提出能够让我们从迷恋单一结构

The proposal of Engineering-Integrated Architecture may promote several changes. One is teaching content and mode. I propose "Fusion of the Fields", which is to break the boundary and let it merge into a big environmental view. In other words, a large construction concept can integrate architecture, urban planning, and landscape architecture into an extensive system for considering problems. The concept of urban landscape infrastructure not only includes infrastructure and architecture, but also becomes a new activity platform in urban public space and a new urban landscape, which can promote changes of teaching content and mode in the future. The second is to promote changes in the way architects and engineers collaborate, and Engineering-Integrated Architecture provides us with a foundation. The third is that the proposal of Engineering-Integrated Architecture can stop us from being obsessed with a single structure to a so-called composite structure. With the combination of technical and art logic, no matter what

向复合性结构转变。当把技术逻辑与艺术逻辑相结合，那么不论什么结构形式对于这两个逻辑而言都是最恰当的，最终都是回到人的体验，回到人对于形态和空间的诉求，在传统的设计体系当中，技术逻辑与艺术逻辑处于一种相对排斥的状态，而工程建筑学的提出，能够把两者从排斥的状态不自觉地转化为整合的状态。所以这三个改变，也是提出工程建筑学所能带来的好的推动。好的工程逻辑不一定能诞生出好的建筑，但是好的建筑一定有好的工程逻辑，它们共同叠加起来就是所谓的工程建筑学。

structural form is the most appropriate for these two, in the end, it is will return to the human experience, back to people's appeal to form and space. In the traditional design system, technical logic and art logic are in a relatively exclusive state, and the proposal of Engineering-Integrated Architecture can unconsciously transform the two from a state of exclusion into one of integration. Therefore, these three changes are also the excellent impetus that Engineering-Integrated Architecture can bring. Good engineering logic may not necessarily produce good buildings, but good buildings must have good engineering logic. They are superimposed together to form the so-called Engineering-Integrated Architecture".

李兴钢：工程建筑学会促进建筑学科和行业的某些有益改变，工程建筑学与技术工程逻辑之间有着深刻的因果关系，离不开对材料和技术的理解及创造性地使用。

罗鹏[1]：工程建筑学的概念强调本真性和原真性的回归。技术本体就是人类认知自然，然后对自然进行抽象，并且作用于自然的一个物质载体或手段，因此技术本体具有强烈的自然性。自然而然，对材料、环境特性、结构逻辑、建造方法的认知，就是对环境建筑要素与技术逻辑的理性认知。技术的自然性主要体现在两个方面，即技术本体的自然性与技术应用的自然性。

[1] 罗鹏：哈尔滨工业大学建筑设计研究院教授、博士生导师。
Luo Peng: associate professor and doctoral supervisor of Architectural Design Institute of Harbin Institute of Technology.

Li Xinggang: Engineering-Integrated Architecture will promote some beneficial modifications in the discipline and industry, and there is a strong causal relationship between Engineering-Integrated Architecture and technical engineering logic, inseparable from the understanding and creative use of materials and technologies.

Luo Peng[1]: Engineering-Integrated Architecture emphasizes the backtracking of authenticity and originality. The ontology of technology is a material carrier or means for human beings to recognize, abstract, from and act on nature. Therefore, the ontology of technology has a strong naturalness. Naturally, the cognition of materials, environmental characteristics, structural logic, and construction methods is the rational cognition of the essential factors of environment and architecture and the technical logic. The naturalness of technology is mainly reflected in two aspects: the naturalness of technology ontology and the naturalness of technology application.

工程建筑学所提到的多学科交叉，不完全是结构和建筑二元对立的内容，而是较大的技术领域，包括材料、结构、建造方法，甚至建造施工的整个过程。建筑的创新不应该把自己限定在空间和形体这个范围内，而应该与各个学科真正协同起来，强化在工程建筑学思想下的协同创新，才能真正推动建筑可持续创新的发展。

　　如何培养具有工程建筑学思想的人。在思想上，让学生建立起整体、系统的建筑观，知道建筑不只是由狭隘的空间和形体组织而构成的，它涉及建筑、建造、生成过程的方方面面；建筑学核心的教育方式就是操作与习得，让学生在真正的建造过程中发现问题、思考问题，才能够真正习得他自己的知识和能力。

　　在工程建筑学的教学过程中，教育应当强化实践的教学方法对于学生们未来发展的作用。

The multidisciplinary intersection mentioned by Engineering-Integrated Architecture is not entirely the binary opposition between structure and architecture but a larger technical field, including materials, structures, detailing methods, and even the entire construction process. The innovation of architecture should not limit itself to the scope of space and form, but instead genuinely cooperate with various disciplines and strengthen the collaborative innovation with the mindset of Engineering-Integrated Architecture to truly promote sustainable architectural innovation.

How to cultivate people with the mindset of Engineering-Integrated Architecture? Ideologically, let students establish a holistic and systematic view of architecture, knowing that architecture is not only composed of parochial spacial and physical organization, but also involves all aspects of architecture, construction, and generation. The core education method of architecture is operation and acquisition. Only when students discover and think about problems in the actual construction process can they truly acquire their knowledge and abilities.

李兴钢：在最前沿、最一线的研究和实践中，不断获得自身对学术和思想方面的认知，这种状态是非常可贵的。所谓的工程建筑学，它所强调的建筑意匠及建筑的诗学传统与工程原理、技术场景之间的交互、统合、触发的一体化思想，其实并不是想要建立一个新的学科来否定传统的建筑学，反而是对传统建筑学、建筑设计还有建筑师职业的某种本质上的回归，或者是对现代以来的传统建筑学所进行的某种拓展和升级。希望它能让我们的建筑学更加具有生命力，更加具有系统性，更加具有科学性，以应对我们的职业工作所服务的人类生存及生活环境的多样的困境和挑战。同时，也希望它能够为我们国家在新的发展时期、发展战略以及全人类发展的背景下能有一个更好的未来，作出一点点可能的贡献。

In the teaching process of Engineering-Integrated Architecture, the role of practical teaching methods should be strengthened for students' future development.

Li Xinggang: In the most pioneering research and practice, continuously acquiring knowledge of academic and ideological aspects is a treasure. The so-called Engineering-Integrated Architecture emphasizes the integration thoughts of interaction, hybridization, and triggering of the architectural ingenuity and poetic tradition of architecture with the engineering principles and technical scenarios. In fact, it is not to establish a new discipline to deny traditional architecture, but in essence to return to the traditional architecture, architectural design, and the profession of the architect, or to continue traditional architecture since modern times with some kind of expansion and upgrade. I hope it could make our architecture more vital, more systematic, and more scientific to cope with diverse dilemmas and challenges of human existence and the living environment served by our professional work. At the same time, I also hope it can make a possible contribution to a better future for our country under the background of the new development period, development strategy, and development of all humanity.

图片来源
参考文献

Figures & Refrences

图片来源

Figures

图 0-1　VITRUVIUS. The Ten Books on Architecture [M]. New York：Dover publications，INC., 1960.

图 0-2　李涵绘制，绘造社提供

图1-1　HELMUT J，WERNER S. Archi-Neering [M]. Germany：Hatje Cantz publishers，1999.

图 1-2　日本建筑学会. 结构建筑创新工学 [M]. 郭屹民，傅艺博，解文静，等译. 上海：同济大学出版社，2015.

图 2-1　李兴钢. 胜景几何论稿 [M]. 杭州：浙江摄影出版社，2020.

图 2-8、图 2-9　结语章末页竖版大图（绩溪博物馆）李哲拍摄

图 2-11　邱涧冰拍摄

图 2-13　是然建筑摄影提供

第三章首图、图 3-9　张虔希拍摄

图 3-1、图 3-3、图 4-5、图 4-12、图 4-13、图 4-14、图 4-17、图 5-9、图 5-12　张广源拍摄

图 3-4、第四章首图　孙海霆拍摄

图 3-11　张音玄拍摄

图 4-8　李季拍摄

图 4-20　李宁拍摄

图 4-23　刘保云拍摄

第五章首图、图 5-6　苏圣亮拍摄

图 5-1　陈颢拍摄

图 5-3、图 5-11　田方方拍摄

图 5-14　祝贺拍摄

除以上特殊注明外，其余所有图、照片均为李兴钢工作室绘制/提供。

参考文献

References

[1]钱锋，余中奇. 结构建筑学——触发本体创新的建筑思维[J]. 建筑师，2015（2）.

[2]斋藤公男. 建筑学的另一种视角——"何为建筑创新工学"[J]. 王西，译. 建筑师. 2015（2）.

[3]李兴钢. 胜景几何论稿[M]. 杭州：浙江摄影出版社，2020.

[4]殷瑞钰，汪应洛，李伯聪，等. 工程哲学[M]. 3版. 北京：高等教育出版社，2018.

[5]李兴钢，郭屹民，等. 建筑学中的工程、技术与意匠[J]. 当代建筑，2021（10）.

图书在版编目（CIP）数据

工程建筑学概论 = Outline of Engineering-
Integrated Architecture：英汉对照 / 李兴钢著. —
北京：中国建筑工业出版社，2022.11
　　ISBN 978-7-112-27996-8

　　Ⅰ.①工… Ⅱ.①李… Ⅲ.①建筑科学—概论—英、
汉 Ⅳ.①TU

　　中国版本图书馆CIP数据核字（2022）第176266号

责任编辑：陆新之　刘　丹
英文校审：尚　晋
责任校对：孙　莹

工程建筑学概论
Outline of Engineering-Integrated Architecture
李兴钢　著
Li Xinggang

*
中国建筑工业出版社出版、发行（北京海淀三里河路9号）
各地新华书店、建筑书店经销
北京锋尚制版有限公司制版
北京雅昌艺术印刷有限公司印刷
*
开本：787毫米×1092毫米　1/16　印张：9　字数：79千字
2023年3月第一版　　2023年3月第一次印刷
定价：**88.00**元
ISBN 978-7-112-27996-8
　　　（40119）